NICOTEXT

...OOPS!

THE TEXT MESSAGES YOU WISH YOU NEVER SENT.

Copyright © NICOTEXT 2012 All rights reserved.
NICOTEXT part of Cladd media ltd.
www.nicotext.com
info@nicotext.com

Printed and bound by CPI Group (UK) Ltd, Croydon, CR0 4YY
ISBN 978-91-86283-10-0

STYLE AND EFFICIENCY

STOP TEXTING ME IF YOU
DON'T WANT TO HAVE SEX.
WON'T BOTHER READING THE
CRAP.

SENT TO: ONE NIGHT STAND

TXT MESSAGE: FLIPPIN' WEIRD. SHE DIDN'T MAKE A SOUND, NOT A MOAN, NOT EVEN BREATHING. I DIDN'T KNOW IF SHE SLEPT OR IF SHE WAS UNCONSCIOUS.

RESPONSE: BECAUSE I DIDN'T FEEL YOU!

MEANT FOR: A BUDDY

EXPLANATION: FUCKING CHICKS. ISN'T IT COMMON COURTESY TO AT LEAST ACKNOWLEDGE THAT I'M ON TOP OF HER?

SENT TO: MY BOSS

TXT MESSAGE: HEY, I LOOKED UP CLINICS THAT DO NOSE JOBS. IF YOU WANT I'LL GO WITH YOU.

RESPONSE: I DON'T WANT A NOSE JOB.

MEANT FOR: MY SISTER

EXPLANATION: I MEANT TO SEND THIS TO MY SISTER WHO HAD BEEN THINKING ABOUT A NOSE JOB FOR A LONG TIME. THE FUNNY THING IS, MY BOSS REALLY HAS A BIG NOSE.

SENT TO: MUMMY DEAREST

TXT MESSAGE: I HAD A DILDO IN MY DRAWER. WHERE THE HELL IS IT? DID YOU TAKE IT?

RESPONSE: ABSOLUTELY NOT! I DON'T HAVE STUFF LIKE THAT.

MEANT FOR: MY BOYFRIEND

EXPLANATION: OOPS, TOO QUICK WITH MY FINGERS THERE. MEANT TO ASK MY BOYFRIEND IF HE TOOK IT...

SENT TO: GIRLFRIEND EMMA

TXT MESSAGE: SORRY, SEXY BABY, CAN'T BE WITH YOU TONIGHT, EMMA THOUGHT WE SHOULD PLAN OUR WEDDING.

RESPONSE: THE WEDDING IS OFF.

MEANT FOR: MY MISTRESS

EXPLANATION: I KNOW I AM A REAL DOUCHE BAG, BUT I HAD A MISTRESS AT THE SAME TIME AS I WAS PLANNING TO GET MARRIED WITH MY GIRLFRIEND. GOD MUST HAVE SEEN THIS AND PUT HIS FINGER ON ME. ANYWAY, THE WEDDING WAS CANCELLED AND NOW I'M LIVING WITH MY MISTRESS.

SENT TO: MY MOTHER

TXT MESSAGE: JUST SO YOU KNOW,
I'M SLEEPING WITH YOUR HUSBAND!

RESPONSE: WITH YOUR FATHER?

MEANT FOR: MY LOVER'S WIFE

EXPLANATION: I WAS SO MAD AT MY LOVER THAT I
WAS TELLING HIS WIFE ABOUT THE TWO OF US, BUT
ACCIDENTALLY SENT IT TO MY MUM INSTEAD.

SENT TO: MY FATHER

TXT MESSAGE: I'M IN YOUR BED WAITING FOR SOMETHING...SOMETHING LONG AND HARD ;)

RESPONSE: ARE YOU AT DAN'S HOUSE?

MEANT FOR: MY BOYFRIEND DAN

EXPLANATION: SMOOTH, DAD SAVED FACE FOR ME THERE.

SENT TO: MUM

TXT MESSAGE: YOU KNOW THAT BAD ASS GUY THAT HIT ON ME AT THE PARTY LAST WEEK? WELL WE HOOKED UP ON MSN AND I GOT HIS NUMBER. GUESS WHO I SLEPT WITH LAST NIGHT? ;)

RESPONSE: THAT'S WHAT I ALWAYS SUSPECTED. BUT DARLING, KNOW THAT YOUR DAD AND I WON'T LOVE YOU ANY LESS BECAUSE YOU ARE A HOMO-SEXUAL.

MEANT FOR: MY BEST GIRLFRIEND

EXPLANATION: GREAT, NOW MY MUM KNOWS I SLEEP WITH GUYS.

SENT TO: TESSA

TXT MESSAGE: TESSA HAS BEEN UP THERE FOR LIKE AN HOUR NOW. CAN'T SHE HEAR THAT SHE SOUNDS LIKE SHE'S GOT A PINEAPPLE UP HER ASS?

RESPONSE: THANKS SO FUCKING MUCH. SCREW YOU, BITCH!

MEANT FOR: MY FRIEND ACROSS THE TABLE

EXPLANATION: WE WERE AT A KARAOKE CLUB BUT TESSA WAS HOGGING THE MIKE. PERHAPS NOW AT LEAST SHE REALIZES SHE'LL NEVER BE AN AMERICAN IDOL WINNER.

SENT TO: MY MENTOR

TXT MESSAGE: HEY! GOT MY MENTOR'S NUMBER TODAY. SHE'S SO HOT, WOULDN'T MIND GIVING HER A CALL AND GETTING HER IN BED ;)

RESPONSE: NO RESPONSE

MEANT FOR: MY FRIEND ANTOINE

EXPLANATION: I'M SUCH A DUMBASS. MY MENTOR NEVER MENTIONED IT AND NEITHER DID I.

SENT TO: FATHER-IN-LAW

TXT MESSAGE: SO DAMN TIRED, HAVE TO SKIP
JOGGING TODAY. WAS UP RIDING ALL NIGHT
LONG ;)

RESPONSE: DEBBIE IS WORKING TODAY SO
SHE CAN'T GO JOGGING EITHER.

MEANT FOR: MY GIRLFRIEND

EXPLANATION: ME AND MY GIRLFRIEND GO JOG-
GING EVERY SATURDAY BUT MY FATHER IN LAW
MUST HAVE FIGURED I HAD MADE PLANS TO GO
JOGGING WITH HIS WIFE. IT WAS ALL WRONG.

SENT TO: AUNT RHONDA

TXT MESSAGE: WHEN CAN I TOUCH YOUR BOOBS? NEXT PARTY PERHAPS?

RESPONSE: SHAME ON YOU. HAVE YOU HIT YOUR HEAD ON SOMETHING?

MEANT FOR: MY COLLEAGUE

EXPLANATION: I HAD THE HOTS FOR MY COLLEAGUE, WHO'D JUST GOTTEN BREAST IMPLANTS, BUT I MIXED UP MY OUTGOING.

SENT TO: GRANDPA

TXT MESSAGE: YOU FAT FUCK. GET OFF OF THAT COUCH AND COME TO THE GYM WITH ME.

RESPONSE: I CAN'T, MY BACK STILL HURTS AFTER THE OPERATION. AND IT'S THE CORTISONE THAT MAKES ME GAIN WEIGHT.

MEANT FOR: MY FRIEND

EXPLANATION: SENT THIS TO MY GRANDFATHER WHO HAD JUST COME HOME FROM SURGERY. HAD TO CALL HIM AND EXPLAIN BUT STILL FELT BAD.

SENT TO: BOSS

TXT MESSAGE: DID YOU FART YESTERDAY? JUST FINISHED WASHING YOUR UNDERWEAR BUT I HAVE TO DO THEM AGAIN TO GET THE RACING STRIPES OFF. LOVE YOU!

RESPONSE: HA-HA! WAS THIS MEANT FOR YOUR HUSBAND? TELL HIM TO TAKE A SHOWER.

MEANT FOR: HUSBAND

EXPLANATION: MY HUSBAND WAS PISSED OFF WHEN I TOLD HIM AND SAID HE WOULD NEVER COME TO OUR OFFICE'S CHRISTMAS PARTY AGAIN.

SENT TO: LAST NIGHT'S DATE

TXT MESSAGE: FELT LIKE I JUST SHAT OUT A PINECONE — WITH THE HOOKS THE WRONG WAY.

RESPONSE: DON'T LIKE THE SOUND OF THAT. WAS SOMETHING WRONG WITH THE FOOD?

MEANT FOR: MY BUDDY

EXPLANATION: HAD INDIAN FOOD WITH A GIRL I AM DATING AND IT MUST HAVE BEEN TOO SPICY OR SOMETHING. HOWEVER, SHE IS, LIKE, THE LAST ONE I WANTED TO TALK TO ABOUT THIS.

SENT TO: OLDER COUSIN

TXT MESSAGE: I'VE LET MY PUSSY HAIR GROW OUT. IT'S LIKE AN AFRO. HA-HA! I'M SUCH A DOPE. WHAT ARE YOU DOING?

RESPONSE: WHAAAT! HA-HA! YES, YOU ARE SUCH A DOPE.

MEANT FOR: SIMON, MY BOYFRIEND

EXPLANATION: MY COUSIN, WHO'S, LIKE, TEN YEARS OLDER, MUST HAVE THOUGHT I WAS CRAZY AND VERY IMMATURE FOR SENDING THIS TO HIM.

SENT TO: MY DAD

TXT MESSAGE: DID YOU SEND ME THE SEXY
UNDERWEAR??? THEN MAYBE YOU CAN WATCH
ME WEAR IT THIS WEEKEND ;)

RESPONSE: IT WASN'T ME, SWEETHEART!

MEANT FOR: A GUY I'VE BEEN DATING

EXPLANATION: I DON'T THINK MY FATHER
WANTS TO SEE ME WEAR SEXY UNDERWEAR...
EMBARRASSING!

SENT TO: AUNT MARIE

TXT MESSAGE: I WAS ALMOST MORE DRUNK THIS
TIME. NOT SURE WHAT THE HELL I DID. HEARD
THAT ME AND MIMI GOT UNDRESSED AND STARTED
MAKING OUT. DON'T REALLY BELIEVE IT BUT I FEEL
LIKE SHIT TODAY.

RESPONSE: I CAN'T SAY I'M HAPPY TO HEAR THIS.
WHY ARE YOU DOING THIS?

MEANT FOR: MY FRIEND MARIE

EXPLANATION: MY AUNT GOT REAL WORRIED AND
CALLED MY MOTHER, WHO CALLED ME AND THE
REST IS GROUNDED-HISTORY...

SENT TO: MY BROTHER

TXT MESSAGE: YOU WANT TO WATCH THE MOVIE WE MADE? IT REALLY TURNED ME ON KNOWING THAT WE WERE FILMING.

RESPONSE: HAVE YOU MADE A PORNO??????

MEANT FOR: MY BOYFRIEND

EXPLANATION: NOT COUNTING MY PARENTS I THINK THE LAST PERSON ON EARTH I WANTED TO KNOW THIS WAS MY BROTHER.

SENT TO: MUM

TXT MESSAGE: AM I A WHORE? CHARLIE LEFT A HUNDRED DOLLAR BILL ON MY NIGHTSTAND WITH A NOTE THAT SAID: FOR YOU, GORGEOUS!

RESPONSE: YOU ARE ABSOLUTELY NOT A WHORE. DON'T SAY SUCH THINGS. HE PROBABLY JUST WANTED YOU TO BUY YOURSELF SOMETHING NICE. KISSES, MUM.

MEANT FOR: MY BEST FRIEND MORGAN

EXPLANATION: OMG. I DIDN'T WANT MY MOTHER TO SEE THIS! NOW SHE NOT ONLY KNOWS THAT I SLEEP AROUND, SHE ALSO KNOWS THAT THIS GUY THINKS I'M A WHORE!

SENT TO: DAD

TXT MESSAGE: I DREAMT I WAS WITH ANOTHER GUY LAST NIGHT. THE WEIRD THING WAS THAT I LIKED IT. TURNED ME ON SO MUCH I HAD TO RUB ONE OUT THIS MORNING. HAVE I LIKE BECOME GAY OVER NIGHT?

RESPONSE: YOU CAN'T HELP WHAT YOU DREAM BUT IF YOU ARE HOMOSEXUAL I WOULD STILL LOVE AND SUPPORT YOU JUST AS MUCH. BUT I HAVE TO ADMIT, IT WOULD FEEL A BIT STRANGE, WITH THE FISHING TRIPS AND ALL. YOU NEED A RIDE TO PRACTICE TONIGHT?

MEANT FOR: JANE, MY BEST GIRLFRIEND

EXPLANATION: IT WAS EMBARRASSING BUT I WAS ACTUALLY HAPPY THAT MY FATHER WOULD BE SO SUPPORTIVE. TURNS OUT, THOUGH, THAT I'M NOT GAY.

SENT TO: MY FATHER

TXT MESSAGE: AY-AY-AYA!! I'M GONNA FUCK TONIGHT!

RESPONSE: YOU ARE COMING HOME RIGHT NOW. I MEAN IT.

MEANT FOR: MY BEST FRIEND, LIBBY

EXPLANATION: THIS IS WHY YOU SHOULD NEVER TEXT WHILE DRUNK.

SENT TO: NEW GIRLFRIEND

TXT MESSAGE: YO YO. JUST GOT HOME FROM KC'S. SHIT HER LEGS ARE HAIRY. LIKE SPIDER LEGS.

RESPONSE: JUST BECAUSE MY LEGS AREN'T AS SMOOTH AS YOURS, YOU FUCKING FAGGOT!

MEANT FOR: MY FRIEND

EXPLANATION: ALL IS NOW FORGOTTEN AND WE ARE ACTUALLY STILL DATING.

SENT TO: BOSS

TXT MESSAGE: IT'S ALL SET. ACTED THE SUFFERING SICK DUDE ALL MORNING AND BOSS THOUGHT I SHOULD GO HOME AND REST. IT'S PARTY!

RESPONSE: SEE YOU AT WORK TOMORROW AT 8. NOT ONE MINUTE LATER.

MEANT FOR: A FRIEND

EXPLANATION: GOING BACK TO WORK THE NEXT MORNING WAS A BIT NERVE WRECKING BUT I THINK MY BOSS HAS SEEN IT ALL BECAUSE HE DIDN'T MENTION ANYTHING.

SENT TO: AUNT VERA

TXT MESSAGE: NOW I CAN DO THE SPLITS. THERE'S JUST A TINY BIT LEFT BUT IF YOU HOLD MY PACKAGE I THINK I CAN DO IT.

RESPONSE: I DON'T THINK THAT'S FUNNY.

MEANT FOR: MY FRIEND

EXPLANATION: ME AND MY FRIEND SEND WEIRD TEXTS TO EACH OTHER, BUT THIS TIME AUNT VERA GOT IN BETWEEN.

SENT TO: MY MOTHER

TXT MESSAGE: GOTCHA. FOUND YOUR VIBRATOR. YOU DIRTY GIRL☺

RESPONSE: I DON'T LIKE THAT YOU LOOK THROUGH MY THINGS. NOW YOU KNOW I HAVE ONE BUT LOTS OF PEOPLE DO AND IT'S PERFECTLY NORMAL.

MEANT FOR: MY GIRLFRIEND

EXPLANATION: I COULD NEVER HAVE IMAGINED THAT MY MUM HAD A VIBRATOR. IT WAS MORE IN-FORMATION THAN I WANTED AND AFTER THIS IT BECAME WEIRD AT HOME FOR A WHILE.

SENT TO: MY STEPDAD

TXT MESSAGE: YOU KNOW THAT MOVIE WHERE THE GIRL GETS JIZZ IN HER HAIR? MAYBE I SHOULD TRY IT. COULD BE LIKE A SECRET RECIPE OR SOMETHING.

RESPONSE: NO I DON'T THINK YOU SHOULD

MEANT FOR: CAROLYN, MY FRIEND

EXPLANATION: I CAN'T BELIEVE I ASKED MY STEP-DAD IF I SHOULD PUT SPERM IN MY HAIR. IT WAS A TOUGH COUPLE OF DAYS BEFORE I COULD FACE HIM AGAIN.

SENT TO: COLLEAGUE

TXT MESSAGE: I'M IN THE BATHROOM. STILL VERY CONSTIPATED. CAN'T GET OUT EVEN THE SMALLEST LITTLE POOP.

RESPONSE: OK, THAT'S A LITTLE MORE INFO THAN I WOULD HAVE WISHED FOR.

MEANT FOR: MY HUSBAND

EXPLANATION: WHEN I GOT THE REPLY I STAYED INSIDE AS LONG AS I COULD, BUT EVENTUALLY I HAD TO FACE MY COLLEAGUE, RED FACED.

SENT TO: ME (THE SON)

TXT MESSAGE: GET UNDRESSED, I'M COMING HOME.
BRINGING WINE AND CHEESE AND MY TONGUE TO
LICK YOU ALL OVER!

RESPONSE: JEEEZUS DAD!!

MEANT FOR: MY MOTHER

EXPLANATION: I DIDN'T REALIZE MY FATHER SENT
THESE KINDS OF MESSAGES TO MY MOTHER. WHAT
THE EFF! I FEEL LIKE USING THAT MEMORY ERASER
THEY HAVE IN MEN IN BLACK.

SENT TO: BOSS

TXT MESSAGE: I FUCKING HATE YOU. YOU ARE SUCH A FUCKING SCUMBAG. GO TO HELL AND DIE!

RESPONSE: NOT SURE WHAT I DID. LET'S TALK ABOUT IT AT WORK TOMORROW.

MEANT FOR: MY BOYFRIEND

EXPLANATION: FOUND OUT MY BOYFRIEND WAS CHEATING ON ME AND THOUGHT I WAS SENDING A HATE MESSAGE TO HIM. APPARENTLY MY BOSS GOT IT INSTEAD. I EXPLAINED IT TO HIM AND HE UNDERSTOOD. AND I GOT RID OF THE BOYFRIEND.

SENT TO: DAD

TXT MESSAGE: 2 PACKS OF MARLBOROS, A 12 PACK OF BUD, A 6 PACK OF SMIRNOFF ICE AND A COUPLE OF SPLIFFS. YOU'LL GET THE MONEY WHEN YOU COME OVER, IF YOU DON'T WANT ANYTHING ELSE...:)

RESPONSE: I'M COMING TO GET YOU RIGHT NOW. ARE YOU AT SOFIE'S? STAND OUTSIDE IN 10 MINUTES OR I WILL COME UP AND EMBARRASS YOU.

MEANT FOR: THIS GUY

EXPLANATION: WE USE THIS OLDER GUY TO BUY ALCOHOL FOR US SOMETIMES AND WHEN MY DAD FOUND OUT HE WAS REALLY PISSED OFF.

SENT TO: MY BOSS

TXT MESSAGE: WOULD YOU BE HAPPY OR MAD IF I TOLD YOU I GOT YOUR NAME TATTOOED RIGHT NEXT TO MY PUSSY?

RESPONSE: WHY WOULD I WANT TO KNOW?
HAVE YOU?

MEANT FOR: MY BOYFRIEND

EXPLANATION: DO I HAVE TO SAY THAT I HAD TO BRACE MYSELF TO GO TO WORK THE NEXT DAY?

SENT TO: MY BROTHER

TXT MESSAGE: I KNOW THIS SOUNDS STRANGE BUT
IF I WANTED YOU TO, WOULD YOU SLEEP WITH ME?
I MEAN IF NOBODY WOULD FIND OUT. IT SOUNDS
WEIRD BUT WOULD YOU?

RESPONSE: THIS MUST BE THE WEIRDEST THING
I'VE EVER HEARD. HAVE YOU THOUGHT ABOUT
THIS A LONG TIME??

MEANT FOR: OUR NEIGHBOR JAKE

EXPLANATION: I HAD THOUGHT ABOUT IT
FOR A LONG TIME, BUT NOT WITH MY BROTHER!
EEWWWWWW!

SENT TO: MY FATHER

TXT MESSAGE: I SAW SOMETHING SICK ON TV THE OTHER DAY. IT WAS ABOUT PORN STARS THAT BLEACHED THEIR ANUSES. WHERE DO THEY GET THAT STUFF?

RESPONSE: HMM, I DON'T KNOW. ASK YOUR MOTHER, SHE KNOWS ALL SORTS OF THINGS.

MEANT FOR: MY FRIEND CLAUDIA

EXPLANATION: THAT WENT TOTALLY WRONG. AND I SHOULD SAY I HAVE NEVER SERIOUSLY CONSIDERED BLEACHING MY ANUS.

SENT TO: MUM

TXT MESSAGE: HEY. WHAT YOU UP TO? I'M PUTTING PINEAPPLE RINGS AROUND MY COCK. YOU WANT TO DO SOMETHING?

RESPONSE: NO RESPONSE

MEANT FOR: GIRLFRIEND

EXPLANATION: MY MUM DIDN'T THINK THIS WAS FUNNY AT ALL. WHERE'S HER SENSE OF HUMOUR — HOW CAN YOU NOT LIKE PINEAPPLE RINGS?

SENT TO: MY BOSS

TXT MESSAGE: NOW I KNOW WHY IT ITCHES SO FUCKING MUCH. CRABS. SOUNDS DISGUSTING, NO?

RESPONSE: WOW, DON'T LIKE THE SOUND OF THAT. SEE YOU AT WORK MONDAY?

MEANT FOR: MY BEST FRIEND

EXPLANATION: I TELL MY BEST FRIEND EVERYTHING. NOW IT SEEMS I ALSO TELL MY BOSS EVERYTHING...

SENT TO: TONIGHT'S DATE

TXT MESSAGE: SEXY. WHEN I TOOK A DUMP TODAY I COULD SEE UNDIGESTED CORN IN IT.

RESPONSE: NO, THAT'S NOT SEXY.

MEANT FOR: A BUDDY

EXPLANATION: SENT THIS TO THE GIRL I WAS GOING ON A DATE WITH LATER THAT NIGHT. I CAN'T THINK OF A WORSE WAY TO BEGIN.

SENT TO: ME

TXT MESSAGE: HI HONEY, JUST WANTED TO SAY THANKS FOR THE BLOWJOB THIS MORNING. YOU ARE FANTASTIC!

RESPONSE: BUT DAAAD! HOW CAN YOU SEND THIS TO ME? I'M DELETING THIS AT ONCE.

MEANT FOR: MY MOTHER

EXPLANATION: FIRST OF ALL, I DON'T WANT TO KNOW WHAT MY MOTHER DOES TO MY FATHER. SECONDLY, I DON'T WANT TO HEAR ABOUT IT. PART OF MY CHILDHOOD IS NOW RUINED.

SENT TO: BIG BROTHER

TXT MESSAGE: HA-HA, WHAT A NIGHT! DROPPED MY BROTHER'S TOOTHBRUSH IN THE TOILET WHEN I GOT HOME. A BIT DRUNKETY-DRUNK. HE'S ALWAYS AT HIS GIRL'S HOUSE ANYWAY SO I WILL HAVE TIME TO CHANGE IT BEFORE:)

RESPONSE: WHAT THE EFF! I SLEPT AT HOME LAST NIGHT! PREPARE FOR PAIN.

MEANT FOR: MY FRIEND

EXPLANATION: I TOLD HIM IT WAS JUST A JOKE AND HE BOUGHT IT. BUT IT WAS REALLY TRUE...

SENT TO: GRANDMA

TXT MESSAGE: IF YOUR BOYFRIEND'S CUM TASTES
LIKE PISS MAKE HIM EAT LOTS OF FRUIT, THAT'S
WHAT I DID WITH MY BF, AND NOW HIS CUM TASTES
GOOD☺

RESPONSE: IS THIS EMMA? SOMETHING IS WRONG
WITH MY PHONE.

MEANT FOR: MY CO-WORKER SARA

EXPLANATION: IT TOOK A FEW HOURS FOR MY
GRANDMA TO GET BACK TO ME. I THINK SHE CHOSE
TO BELIEVE IT WAS FROM SOMEONE ELSE, AND NOT
HER CUM-SWALLOWING GRANDDAUGHTER...

SENT TO: DATE

TXT MESSAGE: SHE MUST HAVE BEEN BLIND AS A BAT OR SOMETHING, KISSING ME WHEN I HAD THAT BIG COLD SORE. MAYBE SHE'S IMMUNE OR SOMETHING. ANYWAY, IT WAS A GREAT KISS.

RESPONSE: EEWWW! WHY DIDN'T YOU TELL ME!? DAMMIT, I LIKE YOU BUT ISN'T THIS SOMETHING YOU SAY BEFORE?

MEANT FOR: MY FRIEND

EXPLANATION: I THOUGHT I KILLED MY CHANCES WITH THIS, BUT WE ARE STILL DATING, AND NOW IT'S A GREAT DINNER CONVERSATION☺

SENT TO: EX-BOYFRIEND

TXT MESSAGE: SORRY FOR EVERYTHING. I LOVE YOU MORE THAN EVER. HOPE YOU CAN FORGIVE ME.

RESPONSE: ARE YOU SERIOUS? I NEVER STOPPED LOVING YOU.

MEANT FOR: CURRENT BOYFRIEND

EXPLANATION: THIS WAS A REAL JERKY MOVE. I HAD LEFT MY EX MONTHS AGO AND HAD A SMALL FIGHT WITH MY NEW BOYFRIEND. I MEANT TO TEXT HIM BUT MY MIND SLIPPED. IT FELT REALLY CRUEL EXPLAINING TO MY EX THAT I JUST MADE A TEXT MISTAKE...

SENT TO: GRANDMA

TXT MESSAGE: HOLY CRAP. I ALMOST THREW UP WHEN I SUCKED JIM LAST NIGHT. HE CAME IN MY MOUTH AND HIT THE LITTLE THING THAT HANGS DOWN IN THE BACK OF MY THROAT. CAN I COME OVER?

RESPONSE: WE AREN'T HOME SWEETHEART. WE'RE IN FLORIDA.

MEANT FOR: MY BEST FRIEND

EXPLANATION: I'M NOT SURE IF GRANDMA GOT IT, OR IF SHE WAS JUST POLITE NOT TO MENTION IT. MAYBE SHE'S BEEN THERE AS WELL.

SENT TO: RECENT DATE

TXT MESSAGE: CONGRATULATIONS! YOU'RE GONNA BE A DAD! I LOVE YOU!

RESPONSE: I'M CALLING YOU RIGHT NOW

MEANT FOR: A FRIEND

EXPLANATION: MY DATE GOT A REAL SHOCKER. HE SAID THE HAIR ON HIS BACK STOOD UP FOR, LIKE, DAYS.

SENT TO: GIRLFRIEND EDINA

TXT MESSAGE: I'M A DIMWIT. SLEPT AT EDINA'S LAST NIGHT AND SHE ASKED ME TO TAKE HER DOG FOR A WALK BEFORE I LEFT. SO I WENT TO THE CORNER STORE AND LEFT, FORGETTING THAT I TIED THE POOR THING TO THE LAMPPOST. NOW I'M ON MY WAY BACK TO GET IT. BUT WHAT IF IT'S NOT THERE? DON'T WANT TO BE HIT BY EDINA'S SHIT STORM.

RESPONSE: IF HE'S NOT THERE I PROMISE YOU WILL REGRET EVER MEETING ME. WAIT FOR ME AT HOME. I'LL BE THERE IN 30. YOU'RE A MORON.

MEANT FOR: MY ROOMMATE

EXPLANATION: LUCKILY, ALL ENDED WELL AS THE DOG WAS STILL THERE. NO HARM DONE, THE DOG PROBABLY NEVER NOTICED ME LEAVING. I ACTUALLY ENDED UP WINNING ON THIS SINCE EDINA HAS NEVER ASKED ME ONCE AGAIN TO WALK HER DOG.

SENT TO: SHIFT SUPERVISOR

TXT MESSAGE: YOU EVER EXPERIENCED THE SECOND
WAVE? YOU KNOW, WHEN YOU SHIT, LEAVE THE TOILET,
AND AFTER A MINUTE OR SO YOU FEEL YOU'RE NOT
DONE? I'M HAVING SECOND WAVE RIGHT NOW BUT
SOME JERK HAS OCCUPIED THE TOILET. WHAT DO
I DO?

RESPONSE: NO, CAN'T SAY I HAVE.
JUST HOLD ON.

MEANT FOR: MY FRIEND

EXPLANATION: DIDN'T MEAN FOR MY FEMALE SHIFT
SUPERVISOR TO LEARN ABOUT MY SHITTING PATTERN.
C'EST LA VIE

SENT TO: GRANDPA

TXT MESSAGE: HE WAS HELL BENT ON GETTING IT IN MY ASS. WHAT THE HELL HAPPENED TO TODAY'S MEN? IF THEY DON'T WANT PUSSY ANYMORE THEY MIGHT AS WELL BECOME GAY!

RESPONSE: I'M WATCHING THE GAME.

MEANT FOR: A CLOSE FRIEND

EXPLANATION: I BLUSH WHEN I THINK ABOUT THIS. POOR GRANDPA. FOR A WHILE I REFUSED TO GO AND SEE MY GRANDPARENTS BECAUSE OF THIS.

SENT TO: MUM

TXT MESSAGE: SWEETIE, I'M DESPERATE! CAN YOU BUY A MORNING AFTER PILL FOR ME? OUT OF CASH AND I CAN'T LET MY MOTHER KNOW.

RESPONSE: THIS DOESN'T SOUND LIKE YOU. WE CAN TALK ABOUT THIS TONIGHT.

MEANT FOR: A FRIEND

EXPLANATION: I DIDN'T LOOK FORWARD TO THAT TALK BUT ACTUALLY AFTERWARDS IT FELT LIKE I COULD TELL HER THINGS, SO THANKS TEXT GODS!

SENT TO: MY GRANDMOTHER

TXT MESSAGE: WENT BY YOUR PLACE THIS MORN-
ING AND SAW THROUGH THE WINDOW WHEN YOU
CHANGED. THERE WAS NO GRASS ON THAT PLAY-
ING FIELD! ;) !

RESPONSE: SHE CALLED.

MEANT FOR: A CO-WORKER I USUALLY JOKE
AROUND WITH.

EXPLANATION: MY GRANDMOTHER WONDERED WHAT
I WAS DOING WHEN SHE CALLED. I REALLY DON'T
WANT TO THINK ABOUT IF SHE HAS ANY GRASS ON
HER PLAYING FIELD. YUCK!

SENT TO: MUM

TXT MESSAGE: I KILLED HIM! NOW WHAT SHOULD I DO?

RESPONSE: SHE CALLED ME THAT VERY SECOND.

MEANT FOR: MY FRIEND JERRY

EXPLANATION: BORROWED JERRY'S GAME BOY AND WAS TOTALLY INTO THIS GAME FOR A COUPLE OF DAYS. BUT I JUST COULDN'T KILL THE LAST MONSTER AND WHEN I DID I HAD TO TELL HIM. MY MUM THOUGHT I HAD KILLED SOMEONE FOR REAL. THE FUNNY THING WAS THAT SHE SAID THAT IF I HAD REALLY KILLED SOMEONE WE'D THINK OF SOMETHING.

SENT TO: GRANDMA

TXT MESSAGE: STAY ON YOUR BACK WHEN YOUR MAN CUMS IN YOU, ALL RIGHT? THAT'S HOW YOU GET PREGNANT.

RESPONSE: NO RESPONSE

MEANT FOR: CO-WORKER JENNIFER

EXPLANATION: MY CO-WORKER WAS TRYING TO GET PREGNANT, AND SHE AND MY GRANDMOTHER HAVE THE SAME FIRST NAME. MY GRANDMOTHER CALLED ME BUT SHE ONLY LAUGHED AND SAID SHE'D CLOSED THE DOOR TO THAT FACTORY A LONG TIME AGO.

SENT TO: MY MUM

TXT MESSAGE: YIPEE-YA-YA! I GOT MY MUM TO SIGN IT. HERE WE COME MIAMI, DRINKS, TOPLESS DANCING, AND SEX!

RESPONSE: YOU ARE GIVING ME BACK THAT NOTE RIGHT AWAY, AND I WILL CALL THE TRAVEL AGENCY AND TELL THEM NOT TO LET YOU GO ANYWHERE UNTIL YOU ARE 18. I'M VERY DISAPPOINTED IN YOU.

MEANT FOR: MY BEST FRIEND ANNA

EXPLANATION: MY MUM WASN'T VERY HAPPY AND NEITHER WAS MY FRIEND ANNA WHO HAD TO GO BY HERSELF.

SENT TO: GIRLFRIEND KATJA

TXT MESSAGE: AT A FAMILY DINNER AND HONESTLY, MY COUSIN IS SO DAMN HOT! IS IT OK TO HAVE THE HOTS FOR YOUR OWN COUSIN?

RESPONSE: YOU ARE A FUCKING RETARD FOR SENDING THIS TO ME. GO AND DIE!

MEANT FOR: MY BUDDY

EXPLANATION: IT TOOK MONTHS TO REPAIR THIS BUT NOW ALL IS WELL AGAIN. BUT MY COUSIN IS UNDENIABLY HOT.

SENT TO: BIG BROTHER

TXT MESSAGE: (PICTURE OF MY BOOBS) BET YOU'D LIKE TO PLAY WITH THESE?

RESPONSE: WHO'S ARE THOSE? I HOPE THEY AREN'T YOURS BECAUSE I JUST JERKED OFF TO THEM.

MEANT FOR: MY BOYFRIEND

EXPLANATION: OBVIOUSLY I TOLD HIM THEY WEREN'T MINE. I'M PRETTY SURE MY BROTHER WAS JOKING ABOUT THE JERKING OFF...

SENT TO: DATE

TXT MESSAGE: NOT SURE ABOUT THIS. INVITED TO
DINNER AT MADELEINE'S HOUSE AND THERE WILL
BE MUSHROOMS IN THE SAUCE. I'VE HEARD THAT
MUSHROOMS WILL PLUG YOUR BUTT HOLE SO YOU
CAN'T TAKE A DUMP.

RESPONSE: JUST CHEW YOUR FOOD PROPERLY
AND YOU SHOULD BE FINE. SEE YOU SOON.

MEANT FOR: MY KNOWLEDGEABLE FRIEND

EXPLANATION: SHE MUST HAVE THOUGHT I WAS A
NUTCASE FOR TEXTING THIS. I THINK I'M A NUT-
CASE EVERY TIME I READ IT. BTW. I NOW KNOW
THAT MUSHROOMS DON'T PLUG YOU UP.

SENT TO: MUM

TXT MESSAGE: GOT ONE OF THOSE VIBRATORS THAT LOOKS LIKE A LIPSTICK, AND MY MUM SAW IT AND ASKED IF SHE COULD USE IT. I PANICKED AND DIDN'T KNOW WHAT TO SAY. SO I JUST LEFT.

RESPONSE: OH, I DIDN'T KNOW.

MEANT FOR: MY BEST FRIEND

EXPLANATION: SOOO EMBARRASSING! I DIDN'T WANT TO GO HOME FROM SCHOOL WITH MY "VIBRATOR" SO I DUMPED IT IN A GARBAGE CAN ON MY WAY HOME FROM SCHOOL.

SENT TO: ONE NIGHT STAND

TXT MESSAGE: WENT HOME WITH THE VIN DIESEL GUY YESTERDAY. HE WENT DOWN ON ME AND I FELT LIKE I HAD TO FART AND TRIED TO HOLD IT BACK, BUT IT SORT OF SLIPPED OUT. HA-HA-HA!

RESPONSE: I NEVER NOTICED BUT I'M PRETTY DISGUSTED NOW THAT I KNOW.

MEANT FOR: MY ROOMMATE SARA

EXPLANATION: EVERYBODY DOES IT. THAT'S WHAT I KEPT TELLING MYSELF AFTER. SARA THOUGHT IT WAS THE FUNNIEST THING SHE'D EVER HEARD.

SENT TO: TEACHER

TXT MESSAGE: I REALLY CAN'T MAKE IT TO SCHOOL TODAY. I'M TOO TIRED. WE HAD SEX LIKE 4 TIMES LAST NIGHT.

RESPONSE: COME TO SCHOOL RIGHT AWAY AND I WON'T MENTION THIS TO YOUR PARENTS!

MEANT FOR: MY FRIEND

EXPLANATION: MY TEACHER AND FRIEND HAVE THE SAME FIRST NAME. NOW MY TEACHER IS UNDER "TEACHER" IN MY PHONE...

SENT TO: BOSS

TXT MESSAGE: I'VE CALLED IN SICK. YOU DO THE SAME AND WE CAN GO TO VEGAS TOMORROW.

RESPONSE: IF YOU ARE NOT BACK AT WORK WITHIN THE HOUR YOU CAN LOOK FOR ANOTHER JOB. PREFERABLY IN VEGAS.

MEANT FOR: MY BUDDY TIM

EXPLANATION: I STILL ENDED UP GOING TO VEGAS. I TEXTED MY BOSS FROM THERE AND SAID I HADN'T FOUND A JOB YET BUT I QUIT ANYWAY. HA!

SENT TO: GRANDMA

TXT MESSAGE: COULD I DUMP SOME SNUFF AT YOUR PLACE?

RESPONSE: NO, I DON'T WANT THAT IN MY HOUSE. WHERE DID YOU GET IT?

MEANT FOR: GRANDMA

EXPLANATION: I MEANT TO WRITE STUFF BUT I MESSED UP THE SPELLING. NOW MY GRANDMA THINKS I'M DOING SNUFF.

SENT TO: RICKY

TXT MESSAGE: I JUST HAD SEX WITH RICKY BUT FORGOT TO TAKE MY TAMPON OUT. BUT HIS DICK SO SMALL HE NEVER NOTICED. NOW THE STRING IS GONE, WHAT THE EFF SHOULD I DO?

RESPONSE: NO RESPONSE

MEANT FOR: MY FRIEND HANNA

EXPLANATION: I WAS IN THE BATHROOM AND SUDDENLY HEARD THE FRONT DOOR SLAM SHUT. BAD ONE.

SENT TO: MY BOSS

TXT MESSAGE: YO! GOTTA TELL YOU SOMETHING.
I THINK MY BOSS IS GOING COMMANDO. I CAN SEE
HIS DICK THOUGH HIS PANTS. MAKES ME HOT. CALL
ME WHEN YOU CAN.

RESPONSE: AREN'T YOU MARRIED? AND WHY ARE
YOU EYEING MY CROTCH?

MEANT FOR: MY LADY FRIEND

EXPLANATION: NOW MY BOSS THINKS I HAVE THE
HOTS FOR HIM, BUT I WAS SIMPLY JUST TALKING
CRAP WITH MY FRIEND BECAUSE I WAS BORED.

SENT TO: DATE

TXT MESSAGE: CAN'T MAKE IT TONIGHT. MEETING THE SHRIMP. YOU KNOW, THE ONE I COULD HARDLY FEEL, IT JUST TICKLED A BIT. TALK 2MORO.

RESPONSE: I CAN'T MAKE IT EITHER BECAUSE I'M SEEING THE SHRIMP BOAT. YOU KNOW, THE ONE THAT COULD FIT AN ENTIRE SHRIMP BOAT. LET'S NOT TALK TOMORROW.

MEANT FOR: MY FRIEND LIBBY

EXPLANATION: IT'S TOO BAD BECAUSE EVEN THOUGH HE WAS SMALL AS A SHRIMP I REALLY LIKED THIS GUY, BUT WE HAVEN'T SEEN EACH OTHER SINCE.

SENT TO: MY MOTHER

TXT MESSAGE: CAUGHT A GLIMPSE OF YOUR ASS THIS MORNING, I'D SAY PRETTY HOT. ;)

RESPONSE: PLEASE STOP THIS.

MEANT FOR: A FRIEND

EXPLANATION: MY MUM DIDN'T BELIEVE ME WHEN I EXPLAINED THIS WAS MEANT FOR A FRIEND.

SENT TO: FEMALE BOSS

TXT MESSAGE: HA-HA, IT'S GREAT HOW THE BUB-
BLES TICKLE WHEN YOU FART IN THE BATHTUB.
THE SMELL ISN'T THAT GREAT THOUGH.

RESPONSE: YEAH... I CAN'T SAY I HAVE MUCH
EXPERIENCE WITH THIS, WE ONLY HAVE A SHOWER
IN OUR HOUSE. I SUGGEST YOU TAKE ONE AFTER
YOUR BATH.

MEANT FOR: MY GIRLFRIEND

EXPLANATION: SLIGHTLY EMBARRASSING BUT
I THOUGHT THAT WAS A COOL ANSWER FROM MY
BOSS.

SENT TO: DAD

TXT MESSAGE: YOU ALWAYS SAY I'M SWEET AS CANDY, BUT THEN HOW COME YOU NEVER LICK ME?

RESPONSE: WHAT IS THIS? YOU STOP THIS RIGHT NOW!

MEANT FOR: A CUTE GUY

EXPLANATION: WAS JUST FLIRTING WITH THIS CUTE GUY BUT I ACCIDENTALLY SENT IT TO MY DAD INSTEAD. POOR DAD. I'M HAPPY I'LL NEVER BE ONE.

SENT TO: MY KID SISTER (9 YEARS OLD)

TXT MESSAGE: THINKING ABOUT YOU SWEETIE. WANT YOU HERE AND NOW SO I CAN TOUCH YOU. I'M IN BED, READY AND WAITING. ;)

RESPONSE: WHAT? SHOULD I COME?

MEANT FOR: MY GIRLFRIEND

EXPLANATION: THIS MADE ME FEEL VERY CREEPY. MY GIRLFRIEND AND SISTER ARE RIGHT NEXT TO EACH OTHER IN MY CONTACTS.

SENT TO: ME

TXT MESSAGE: WELL, THAT'S WHAT HAPPENS WHEN
YOU ARE YOUNG. I LOST MY VIRGINITY AT 12.
TERRIBLE BUT TRUE! SO DON'T WORRY.

RESPONSE: BUT MUM!

MEANT FOR: MY MUM'S FRIEND

EXPLANATION: MY MUM'S FRIEND WAS WORRIED
ABOUT HER WILD DAUGHTER, WHICH IS WHAT I
GUESS MY MUM USED TO BE...

SENT TO: MUM

TXT MESSAGE: HELP, I CAN'T REMEMBER ANYTHING!
THEY SAY I WAS CRAZY DRUNK AND HAD SEX WITH
PETE. IS THAT TRUE?

RESPONSE: FIRST OF ALL, YOU ARE GROUNDED FOR
THE REST OF YOUR LIFE. AND WHERE DID YOU GET
THE ALCOHOL FROM? YOU ARE IN BIG TROUBLE.

MEANT FOR: A FRIEND FROM THE PARTY

EXPLANATION: I TRIED TO BLAME IT ON A DARE,
THAT THIS WAS JUST ONE OF THOSE TEXTS YOU
SEND TO YOUR PARENTS AS A JOKE. BUT MY MOTHER
DIDN'T FALL FOR IT.

SENT TO: BIG BOSS

TXT MESSAGE: CAN YOU CLEAR THE SNOW FROM OUR DRIVEWAY IF IT SNOWS THIS WEEKEND? WE'RE COMING HOME SUNDAY. COME BY FOR DINNER.

RESPONSE: I'M AFRAID I HAVE THINGS TO DO THIS WEEKEND. WHERE ARE YOU GOING? HAVE A NICE WEEKEND.

MEANT FOR: OUR SON

EXPLANATION: I DIDN'T REALIZE I HAD SENT IT TO THE WRONG PERSON UNTIL THE WEEK AFTER WHEN MY BOSS SAID IT WAS LUCKY THAT IT DIDN'T SNOW THAT WEEKEND.

SENT TO: MY MUM

TXT MESSAGE: I WAS SHOCKED WHEN HE DROPPED
HIS PANTS. HUNG LIKE A DINOSAUR. YOU INVITING
HIM TO THE PARTY SATURDAY, EH?

RESPONSE: WHAT PARTY? IS THIS WHAT YOU DO
WHEN WE ARE AWAY? NEXT TIME YOU ARE COMING
WITH US. YOU ARE GROUNDED UNTIL YOU LEAVE
FOR COLLEGE.

MEANT FOR: MY FRIEND WHO WAS THROWING
A PARTY

EXPLANATION: MY PARENTS WERE AWAY SOME-
WHERE AND MY MOTHER HAS THIS PICTURE OF ME
AS THIS INNOCENT GIRL PLAYING WITH DOLLS OR
SOMETHING, WAITING FOR THEM TO COME HOME.

SENT TO: MY FATHER

TXT MESSAGE: YOU HAVE NO IDEA HOW FUCKING WASTED I WAS THIS WEEKEND. THEY FOUND ME WEARING ONLY PANTIES IN ONE OF THE ROOMS, NOT SURE WHAT HAPPENED. I GOTTA STRAIGHTEN OUT.

RESPONSE: I JUST WANT TO KNOW WHERE YOU GOT THE ALCOHOL FROM.

MEANT FOR: MY FRIEND JOSH

EXPLANATION: I HAD TO TELL MY DAD THAT WE GOT THE BOOZE FROM MY FRIEND'S MOTHER, AND HE CALLED HER UP AND THREATENED TO TURN HER IN. AND ALL THIS BECAUSE OF A TEXT.

SENT TO: MY BOSS

TXT MESSAGE: I LAUGH EVERY TIME I THINK ABOUT IT. CAN'T GET THAT PICTURE OUT OF MY HEAD OF WHEN YOU PUT ON MY PANTIES AND BRA. HAHA! WHAT IF SOMEONE KNEW!

RESPONSE: HA-HA-HA! DID JOHN WEAR THAT YESTERDAY? HA-HA-HA, THIS IS ONE FOR THE BAR!

MEANT FOR: MY HUSBAND

EXPLANATION: MY HUSBAND WAS ONLY TRYING TO MAKE ME LAUGH THE OTHER NIGHT, AND HE CAME OUT OF THE BATHROOM WEARING MY UNDERWEAR. NOW, UNFORTUNATELY, THANKS TO MY FUMBLE FINGERS, HE'S KNOWN AS THE 'THE MANVESTITE' AT HIS LOCAL BAR.

SENT TO: DAD

TXT MESSAGE: I MIGHT JUST BREAK UP WITH MY BOYFRIEND OTHERWISE ;)

RESPONSE: I THINK YOU SHOULD HAVE A LONG TIME AGO. GOOD TO HEAR YOU FINALLY CAME BACK TO YOUR SENSES AGAIN.

MEANT FOR: MY BOYFRIEND

EXPLANATION: I WAS JUST MESSING AROUND WITH MY BOYFRIEND ABOUT HIM NOT FORGETTING TO BUY ICE CREAM ON HIS WAY HOME, BUT NOW I FOUND OUT WHAT MY DAD REALLY THOUGHT OF HIM...

SENT TO: ME

TXT MESSAGE: HEY DARLING! I'M GOING TO BE WITH MY GF TONIGHT SO DON'T CALL ME. WILL CALL YOU WHEN I'M HOME LATER TONIGHT.

RESPONSE: OK, THEN IT'S WHAT I THOUGHT. IT'S OVER!

MEANT FOR: SOMEONE CALLED DARLING...

EXPLANATION: I HAD A HUNCH THAT MY BOY-FRIEND WAS CHEATING ON ME, AND WITH THIS I FINALLY KNEW.

SENT TO: THIS GUY

TXT MESSAGE: OMG. HE'S CALLED ME LIKE 10 TEN TIMES NOW AND I REALLY DON'T WANT TO SPEAK TO HIM.

RESPONSE: I GUESS I'LL STOP CALLING THEN.

MEANT FOR: MY BEST FRIEND

EXPLANATION: HE DID STOP CALLING BUT I FELT BAD FOR TEXTING HIM.

SENT TO: BIOLOGY TEACHER

TXT MESSAGE: HOW DO YOU FEEL ABOUT ME?
I WANT US TO BE TOGETHER.

RESPONSE: NO, NO, THAT'S NOT HOW I FEEL. IT'S
IMPORTANT THAT THIS TEXT DOESN'T REACH ANYONE
ELSE. IT COULD BE INTERPRETED THE WRONG WAY.
SEE YOU THURSDAY.

MEANT FOR: A GUY IN MY BIOLOGY CLASS

EXPLANATION: THEY HAVE THE SAME FIRST NAME
BUT I DON'T REALLY KNOW WHY I EVEN HAVE MY
TEACHER'S NUMBER.

SENT TO: MY MOTHER

TXT MESSAGE: I'M SO FUCKING DISGUSTED. I FOUND A DILDO IN MUM'S DRAWER. WHAT THE HELL DOES SHE USE THAT FOR?

RESPONSE: NO RESPONSE

MEANT FOR: MY BEST FRIEND DARLA

EXPLANATION: I WAS LOOKING FOR SOCKS WHEN I FOUND IT. WE NEVER TALKED ABOUT IT THOUGH. IT'S LIKE A PIECE OF HISTORY THAT DOESN'T EXIST.

SENT TO: MY BOSS

TXT MESSAGE: I AM INCREASINGLY WORRIED. MY LEFT BALL IS HANGING MUCH LOWER DOWN AND IS TWICE THE SIZE. IT DOESN'T HURT BUT I THINK I SHOULD GO SEE SOMEONE.

RESPONSE: YES, I AGREE. I THOUGHT YOU WERE HOME FOR OTHER REASONS BUT NOW I KNOW. COME BACK WHEN YOU ARE BETTER.

MEANT FOR: MY GIRLFRIEND

EXPLANATION: I SENT THIS BY MISTAKE TO MY FEMALE BOSS. SHE PROVED TO BE VERY UNDER-STANDING ABOUT MY BALLS. MY GIRLFRIEND COULDN'T STOP LAUGHING. BTW, MY BALL IS FINE NOW.

SENT TO: ONE NIGHT STAND

TXT MESSAGE: ON MY WAY HOME NOW. SHE HAD THE BIGGEST BUSH I'VE EVER SEEN. I WENT DOWN BUT ALMOST DIDN'T MAKE IT BACK UP. GAVE THE WRONG NUMBER SO I NEVER HAVE TO WALK THROUGH THAT JUNGLE AGAIN. PIZZA?

RESPONSE: THANKS, NOW I HAVE YOUR NUMBER. I'M JUST GOING TO SHAVE, THEN I'LL MEET YOU FOR PIZZA.

MEANT FOR: MY BUDDY JOSH

EXPLANATION: TOTALLY MEAN AND EMBARRASSING, BUT WE'VE ACTUALLY BEEN TOGETHER FOR OVER 2 YEARS NOW.

SENT TO: ME (ELLEN)

TXT MESSAGE: TALKED TO ELLEN AND SHE WILL BE SLEEPING AT CHRISTY'S. THAT MEANS WE HAVE THE ENTIRE HOUSE TO OURSELVES. BUY SOME WINE AND I'LL FIX FOOD, THEN WE'LL SHOWER AND HAVE SEX ALL NIGHT.

RESPONSE: NO, NO! I DON'T WANT TO KNOW THIS!

MEANT FOR: MY FATHER

EXPLANATION: MY MUM SENT THIS TO ME INSTEAD OF MY FATHER. I REALLY DIDN'T WANT TO KNOW WHAT THEY DO WHEN I'M NOT THERE.

SENT TO: CURRENT GIRLFRIEND

TXT MESSAGE: HAD SEX WITH VICKY LAST NIGHT! FEEL LIKE A JERK, BUT A GOLDEN JERK. WHAT SHOULD I DO? JUST PRETEND NOTHING HAPPENED?

RESPONSE: IT'S OVER!!

MEANT FOR: MY ROOMMATE

EXPLANATION: OBVIOUSLY THIS ENDED IN A FLASH.

SENT TO: GRANDFATHER

TXT MESSAGE: YOU WANT TO COME OVER FOR A NIGHTCAP?

RESPONSE: NO SON, WE GO TO BED AT 9 HERE.

MEANT FOR: A GIRL I WAS DATING.

EXPLANATION: GRAMPS IS IN A RETIREMENT HOME. DIDN'T MEAN TO MESS WITH HIS ROUTINE.

SENT TO: MY TEACHER

TXT MESSAGE: HEY, STINKY BALLS! YOU WANT TO CHECK OUT A MOVIE TONIGHT?

RESPONSE: I DON'T THINK THIS IS APPROPRIATE.

MEANT FOR: MY GUY FRIEND

EXPLANATION: I SERIOUSLY CONSIDERED DROPPING OUT OF SCHOOL WHEN I GOT MY TEACHER'S RESPONSE.

SENT TO: RACHEL

TXT MESSAGE: RACHEL IS PREGNANT!!

RESPONSE: I'M GUESSING THIS WAS FOR SOMEONE ELSE? I ASKED YOU NOT TO TELL ANYONE.

MEANT FOR: MY FRIEND KELLY

EXPLANATION: RACHEL HAD ASKED ME NOT TO TELL ANYONE BUT AS SOON AS I LEFT HER PLACE I TEXTED KELLY, OR SO I THOUGHT....

SENT TO: PAMELA

TXT MESSAGE: HAVE YOU SEEN PAMELA'S BABY?
IT ALMOST FRIGHTENED ME. YOU COULD SEE THE
VEINS THROUGH ITS SKIN. A BIT LIKE AN ALIEN.
AND HE LOOKED LIKE HIS FATHER WHICH ISN'T
REALLY IN HIS FAVOUR.

RESPONSE: YOU ARE A REALLY BAD FRIEND.

MEANT FOR: A MUTUAL FRIEND

EXPLANATION: I'M SUCH A DUMBASS.

SENT TO: GRANDMA JANE

TXT MESSAGE: HEY! REMEMBER WHEN WE DROVE TO THE WATER PARK AND I STUCK MY ASS OUT THE WINDOW WHILE YOU LOOKED OUT FROM THE FRONT, AND PEOPLE THOUGHT WE WERE TWINS!

RESPONSE: NO RESPONSE

MEANT FOR: A FRIEND

EXPLANATION: POOR GRANDMA, SHE ASKED ME TO EXPLAIN HOW TO USE HER PHONE NEXT TIME WE MET. SHE SAID SHE WAS GETTING THESE MESSAGES THAT SHE HADN'T ORDERED.

SENT TO: DAD

TXT MESSAGE: OH MAN, IT STINGS AND ITCHES. IS IT LIKE THIS FOR EVERYONE WITH SHAVED BALLS?

RESPONSE: I DON'T HAVE ANY EXPERIENCE WITH THESE SORTS OF THINGS. LOOK ONLINE TO SEE IF YOU CAN DO IT DIFFERENTLY.

MEANT FOR: MY BROTHER

EXPLANATION: SENT THIS TO THE WRONG PERSON. IT SEEMS LIKE THEY DIDN'T SHAVE BALLS IN THE 60'S.

SENT TO: BIG SISTER

TXT MESSAGE: OMG! LOG INTO MSN RIGHT NOW, GOTTA SHOW U A PICTURE OF A ZIT I GOT ON MY DICK. HA-HA!

RESPONSE: YOU ARE NOT FUCKING SENDING THAT TO ME!!!

MEANT FOR: GIRLFRIEND

EXPLANATION: MY SISTER WAS PRETTY GROSSED OUT, BUT I KNOW MY GIRLFRIEND WOULD HAVE APPRECIATED IT.

SENT TO: MY MUM

TXT MESSAGE: LOOKS LIKE MY DICK SHRUNK!

RESPONSE: BUT DEAREST! WHAT'S GOING ON?

MEANT FOR: MY GIRLFRIEND

EXPLANATION: MY MUM IS ON A NO-NEED-TO-KNOW BASIS WHEN IT COMES TO MY DICK SIZE.

SENT TO: GIRLFRIEND JULIA'S KID SISTER

TXT MESSAGE: CAN YOU SHOW YOUR SEXY BODY IN THE WEBCAM? I WANT TO BE ALL OVER YOU.

RESPONSE: YOU ARE GOING TO DELETE THIS NUMBER AND NEVER EVER GET IN TOUCH AGAIN. IF YOU DO, I WILL RIP YOUR BALLS OFF AND BRING THEM OVER TO YOUR PARENTS IN A DIRTY NAPKIN. —KAYLA

MEANT FOR: JULIA

EXPLANATION: JULIA'S FATHER THOUGHT I ACTU-ALLY MEANT TO SEND THIS TO JULIA'S KID SIS-TER, WHO IS ONLY 12 YEARS OLD. I STILL FEEL NERVOUS VISITING THEIR HOUSE.

SENT TO: GIRLFRIEND LUCY

TXT MESSAGE: FAMILY DINNER AT LUCY'S. DYING
OF BOREDOM. PLS CALL AND PRETEND THERE'S AN
OFFICE EMERGENCY SO I CAN LEAVE.

RESPONSE: YOU CAN LEAVE WITHOUT LYING.

MEANT FOR: MY CO-WORKER

EXPLANATION: TRIED EXPLAINING IT,
BUT MY GIRLFRIEND WAS PISSED OFF.

SENT TO: GRANDPA

TXT MESSAGE: YOU AWAKE? CAN I COME OVER AND TALK, PERHAPS WATCH A MOVIE?

RESPONSE: NO RESPONSE

MEANT FOR: A FRIEND

EXPLANATION: I SENT THIS AT MIDNIGHT AND THE NEXT DAY MY GRANDFATHER CALLED AND ASKED IF SOMETHING WAS WRONG.

SENT TO: EX-BOSS

TXT MESSAGE: NOW IT'S FUCKING ABOUT FUCKING TIME THAT I QUIT THIS FUCKING JOB. CAN'T TAKE ANOTHER FUCKING MINUTE. IF YOU HEAR ANYTHING, LET ME THE FUCK KNOW.

RESPONSE: I THINK YOU ARE RIGHT.

MEANT FOR: MY BUDDY

EXPLANATION: MY BOSS ACTUALLY AGREED WITH ME AND GAVE ME A MONTH OFF WITH PAY.

SENT TO: MY DATE

TXT MESSAGE: ON MY WAY TO HIM RIGHT NOW. KEEP YOUR FINGERS CROSSED. HOPE HE DOESN'T WANT TO KISS ME BECAUSE I HAVE THIS COLD SORE OR SOMETHING ON THE INSIDE OF MY LIP.

RESPONSE: UH...WELL NOW I DON'T REALLY WANT TO.

MEANT FOR: A FRIEND

EXPLANATION: I WANTED TO CANCEL EVERYTHING WHEN I GOT HIS ANSWER, BUT IT TURNED OUT TO BE A VERY NICE DATE AFTER ALL.

SENT TO: MY MOTHER

TXT MESSAGE: REALLY HAVE TO DO THAT AGAIN! FELT SO GOOD!

RESPONSE: I'D LOVE TO GO TO THE MOVIES WITH YOU AGAIN.

MEANT FOR: MY BOYFRIEND

EXPLANATION: MY BOYFRIEND WENT DOWN ON ME BUT I ACCIDENTALLY SENT A TEXT TO MY MOTHER WHO THOUGHT I WAS TALKING ABOUT WHEN WE WENT TO THE MOVIES.

SENT TO: MY FORMER BOSS

TXT MESSAGE: MY BOSS IS JUST A PAIN! HE KEEPS PRESSURING ME AND HE'S A PERV THAT ONLY WANTS TO GET INTO MY PANTS. SAW HIM AT JEWELS AT CLOSING, TRYING TO CATCH THEM COMING OUT. FUCKING OLD TIMER, I HATE MY JOB!

RESPONSE: IS THIS A BAD JOKE? FIRST OF ALL I'M NOT A PERV AND SECONDLY I HAVE NO INTEREST IN SLEEPING WITH YOU. WELCOME BACK TO THE OFFICE WHEN YOU CAN FACE ME.

MEANT FOR: KATE, MY ROOMMATE

EXPLANATION: THERE WAS NOTHING I COULD SAY SO I JUST KEPT UP APPEARANCES FOR THE YEAR I STAYED THERE.

SENT TO: BOYFRIEND CRAIG

TXT MESSAGE: ROBERT PATTINSON IS THE SHIT. WOULD SERIOUSLY DO ANYTHING FOR HIM. SOMETIMES WHEN CRAIG AND ME HAVE SEX I FANTASIZE THAT HE IS PATTINSON. IT'S SICK BUT WHAT CAN I DO!

RESPONSE: OK, I SOMETIMES IMAGINE YOU ARE KRISTEN STEWART.

MEANT FOR: TO MY FRIEND EVE

EXPLANATION: WE ACTUALLY HAD A FIGHT ABOUT THIS TEXT, AND LATER BROKE UP.

SENT TO: MUM

TXT MESSAGE: I'M NOT WEARING ANYTHING ;)
WANNA COME OVER AND PLAY DIRTY?

RESPONSE: WHAT ARE YOU DOING?! YOU ARE
SUPPOSED TO BE STUDYING!

MEANT FOR: MY FUCK BUDDY

EXPLANATION: HAD TO EXPLAIN TO MY MUM SEV-
ERAL TIMES THAT I DON'T BRING BOYS BACK HOME
AND HAVE SEX WITH THEM, BUT I DO OF COURSE...

SENT TO: FEMALE BOSS

TXT MESSAGE: I'VE SHIT LIKE AN ANIMAL ALL DAY.
I'M LEAVING NOW.

RESPONSE: SORRY TO HEAR THAT, COME BACK
WHEN YOU FEEL BETTER.

MEANT FOR: MY WIFE

EXPLANATION: FIRST OF ALL, I DIDN'T MEAN TO
WRITE "SHIT," BUT "WORKED," AND I DEFINITELY
DID NOT MEAN TO SEND IT TO MY BOSS. NOT MANY
RIGHTS IN THAT WRONG.

SENT TO: KID SISTER

TXT MESSAGE: ...THEN I BOUGHT A DILDO AND SOME BABY OIL. YOU KNOW, I LIKE THE DIRTY STUFF. I HOPE HE WILL TOO!

RESPONSE: YOU KNOW I'LL BE HOME FRIDAY TOO?

MEANT FOR: A FRIEND

EXPLANATION: MY KID SISTER THOUGHT SHE HAD TO BE IN THE HOUSE WHEN ME AND MY BOYFRIEND HAD DIRTY SEX, POOR THING.

SENT TO: MY GRANDMOTHER

TXT MESSAGE: ARE YOU ALIVE?

RESPONSE: NO RESPONSE

MEANT FOR: A FRIEND FROM HIGH SCHOOL

EXPLANATION: HADN'T HEARD FROM A FRIEND IN A WHILE SO I SENT HIM A TEXT, I THOUGHT... MY GRANDMOTHER TOOK IT PRETTY BAD — SHE THOUGHT I WAS WAITING FOR HER TO DIE SO I COULD GET HER CAR.

SENT TO: SCIENCE TEACHER

TXT MESSAGE: WHAT'S UP? WANNA COME OVER FOR A QUICKIE?

RESPONSE: WHO IS THIS?

MEANT FOR: MY BOYFRIEND

EXPLANATION: MY TEACHER FOUND OUT I SENT THE TEXT AND AFTER THAT HE BECAME REALLY CREEPY.

SENT TO: MY MUM

TXT MESSAGE: MUM'S MET THIS SUPER TASTY HUNK OF A GUY THAT I WOULDN'T MIND FOOLING AROUND WITH.

RESPONSE: I'M HAPPY THAT YOU LIKE GEORGE BUT I HOPE IT'S IN THE RIGHT WAY.

MEANT FOR: MY FRIEND

EXPLANATION: WE NEVER TALKED ABOUT IT.

SENT TO: A TEACHER COLLEAGUE

TXT MESSAGE: COULDN'T FIND ANY CLEAN UNDER-
WEAR THIS MORNING SO I'M WEARING A PAIR OF
YOUR PANTIES☺ WHAT IF SOMEONE KNEW☺

RESPONSE: NOW I KNOW!

MEANT FOR: MY WIFE

EXPLANATION: LUCKILY MY COLLEAGUE KEPT IT
TO HIMSELF, BUT WE BOTH HAD A GOOD LAUGH.

SENT TO: FEMALE BOSS

TXT MESSAGE: HEY SEX MACHINE! YOU WANT TO CATCH DINNER TONIGHT? FOLLOWED BY ME ON TOP OF/INSIDE YOU?

RESPONSE: HELLO! SORRY, HAVE TO SAY NO, BUT THANKS FOR ASKING ;)!

MEANT FOR: MY GIRLFRIEND

EXPLANATION: MY GIRLFRIEND THOUGHT THIS WAS HILARIOUS. ME, NOT SO MUCH...

SENT TO: DANIELLE

TXT MESSAGE: JEFF YOU HAVE TO HELP! YOU'RE A DOCTOR OR SOMETHING, RIGHT? I'M AT DANIELLE'S PLACE AND JUST DROPPED HER GUINEA PIG ON THE KITCHEN FLOOR. IT'S NOT MOVING AND I'M AFRAID TO TOUCH IT. SHE'LL BE BACK SOON. WHAT DO I DO?

RESPONSE: HA-HA, THAT'S FUNNY!!

MEANT FOR: MY FRIEND JEFF

EXPLANATION: SO THE GUINEA PIG REALLY DIED AND I FELT TERRIBLE. IT'S IRONIC THAT DANIELLE GOT THE MESSAGE BY MISTAKE AND THEN THOUGHT I WAS ONLY JOKING. AND IT TURNS OUT JEFF WASN'T A DOCTOR BUT IN MEDICAL SALES...

SENT TO: MY 90-YEAR OLD MOTHER

TXT MESSAGE: I FOUND A PACKAGE OF CIGARETTES IN YOUR BEDROOM. IF YOU ARE SMOKING FORGET ABOUT YOUR CAR.

RESPONSE: SHE CALLED ME

MEANT FOR: MY DAUGHTER JENNIE

EXPLANATION: MY MOTHER, JENNIE'S GRAND-MOTHER, CALLED ME BACK AND SAID SHE INTEND-ED TO KEEP SMOKING, NO MATTER WHAT ANYONE SAID, AND THAT SHE DIDN'T CARE ABOUT THE CAR BECAUSE SHE HADN'T BEEN DRIVING MUCH AFTER SHE TURNED 80 AND STARTED LOSING HER VISION.

SENT TO: DAD

TXT MESSAGE: CAN YOU BUY A CUCUMBER?

RESPONSE: HOW LONG?

MEANT FOR: DAD

EXPLANATION: I SENT THIS FROM MY MOTHER'S PHONE. WHEN MY DAD GOT HOME HE WAS EMBARRASSED AND SAID HE MEANT TO WRITE "HOW MANY?". YEAH, RIGHT. YUCK!!

SENT TO: MY FATHER

TXT MESSAGE: WHEE!! I JUST GOT DICKED LIKE 20 FEET IN THE AIR!

RESPONSE: OK

MEANT FOR: FRIEND JULIE

EXPLANATION: TWO THINGS WENT WRONG. I MEANT TO WRITE KICKED, LIKE ON ONE OF THOSE BUNGY THINGS, AND I WASN'T PLANNING TO SEND IT TO MY DAD. NOW HE THINKS I GOT DICKED BY SOME- ONE. I DON'T EVEN KNOW WHAT THAT MEANS, BUT IT SOUNDS BAD!

SENT TO: MATT

TXT MESSAGE: HE-HE-HE, I JUST TOLD MATT I ONLY HAD 5 BEFORE HIM. I THINK IF I TOLD HIM ABOUT THE 15 HE'D PUT ME IN THE SLUT CATEGORY, AND I ENJOY BEING INNOCENT.

RESPONSE: GOOD THING WE NEVER HAD SEX THEN.

MEANT FOR: MY GIRLFRIEND SHELLY

EXPLANATION: IT'S NOT LIKE I'M A TOTAL SLUT OR ANYTHING, BUT HE DIDN'T WANT TO TALK TO ME AGAIN.

SENT TO: BOYFRIEND LUKAS

TXT MESSAGE: ...YOU KNOW HIS TWIN BROTHER IS EVEN MORE GOOD LOOKING. PERHAPS I SHOULD PULL THE OLD BABY SWITCH!

RESPONSE: WHAT THE EFF? ARE YOU SERIOUS?

MEANT FOR: MY FRIEND

EXPLANATION: I DON'T EVEN KNOW WHY I SENT THIS, BUT AFTER EXPLAINING WE WERE OK AGAIN.

SENT TO: BRETT

TXT MESSAGE: WOULD HAVE BEEN SO NICE TO GET DOWN IN THE JACUZZI WITH YOU...

RESPONSE: BUT I DON'T HAVE A JACUZZI!!!!

MEANT FOR: ANOTHER GUY

EXPLANATION: I WAS DATING THESE TWO GUYS BUT APPARENTLY HAD TROUBLE TELLING THEM APART.

SENT TO: BOSS

TXT MESSAGE: YOU UGLY FUCKING IDIOT, YOU HAVE RUINED MY LIFE! I HOPE YOU CHOKE ON A FUCKING PUSSY AND DIE!

RESPONSE: YOU SOUND UPSET. IF YOU FEEL LIKE IT WE CAN SCHEDULE A TIME WITH THE COMPANY PSYCHOLOGIST?

MEANT FOR: MY HUSBAND

EXPLANATION: HAD JUST FOUND OUT MY HUSBAND WAS CHEATING ON ME BUT, I ACCIDENTALLY SENT IT TO MY BOSS INSTEAD.

SENT TO: MY MOTHER

TXT MESSAGE: MY ONE BALL IS SWOLLEN AND ITCHES LIKE CRAZY. YOU KNOW ANYTHING ABOUT THIS?

RESPONSE: I CAN'T SAY I DO HONEY, BUT IT SOUNDS UNCOMFORTABLE. ASK YOUR FATHER. HE HAD SOME PROBLEMS WITH HIS BALLS ONCE.

MEANT FOR: MY BIG BROTHER

EXPLANATION: MY MUM STILL ASKS ME HOW MY BALL IS DOING...

SENT TO: DAD

TXT MESSAGE: YOU WANT TO DO THIS
TONIGHT? (PICTURE OF MY ASS)

RESPONSE: DON'T WANT YOU TO HAVE
COMPANY OVER TONIGHT. CALLING MUM.

MEANT FOR: MY BOYFRIEND.

EXPLANATION: MY DAD WAS JUST AS
EMBARRASSED AS ME WHEN WE MET.

SENT TO: MY BOSS JIM

TXT MESSAGE: YO NIGGA! LET'S DO LUNCH TODAY AT 12?

RESPONSE: HELLO. SORRY BUT I AM BUSY TODAY. JIM

MEANT FOR: MY BEST BUDDY

EXPLANATION: I WAS PRETTY EMBARRASSED, NOT ONLY FOR SENDING IT TO MY BOSS BUT ALSO DOING SO USING THE "N" WORD.

SENT TO: AARON

TXT MESSAGE: OMG! AARON HAS CHICAGO'S MOST HAIRY BODY, HE LOOKS LIKE A HOMELESS MONKEY! SO HAIRY, SO SCARY...

RESPONSE: YOU CAN DRIVE ME BACK TO THE BUS STATION NOW.

MEANT FOR: MY FRIEND

EXPLANATION: HAD THIS INTERNET FLING WITH A GUY AND WHEN I FINALLY MEET HIM HE HAD THIS ANIMALISTIC HAIR GROWTH ALL OVER HIS BODY. IT WAS PRETTY TENSE IN THE CAR ON THE WAY BACK, TO SAY THE LEAST...

SENT TO: MY HUSBAND

TXT MESSAGE: DON'T YOU THINK TONY IS GETTING ON THE VERGE OF CHUBBY? PERHAPS I'LL GET HIM A GYM CARD FOR HIS B-DAY. WHEN ARE YOU COMING OVER?

RESPONSE: UM, ARE YOU SERIOUSLY TALKING ABOUT MY CHUBBINESS WITH YOUR GIRLFRIENDS?

MEANT FOR: MY GIRLFRIEND

EXPLANATION: I HAD TO APOLOGIZE BUT THE THING IS, HE WAS REALLY GETTING CHUBBY...

SENT TO: GEORGE

TXT MESSAGE: WENT TO GEORGE'S STRAIGHT AF-
TER. THE WORST SEX EVER. I HARDLY FELT ANY-
THING AND EVENTUALLY I JUST FAKED ONE. WHEN
DID YOU LEAVE?

RESPONSE: OK, NOW I KNOW. BUT I THINK THE
WORST IS THAT YOU WERE SENDING THIS TO SOME-
ONE ELSE.

MEANT FOR: MY FRIEND LIBBY

EXPLANATION: WENT HOME WITH THIS GUY IN OUR
GROUP OF FRIENDS BUT HE JUST DIDN'T DO IT FOR
ME. WHAT CAN I SAY?

SENT TO: THE BOSS

TXT MESSAGE: I MIGHT HAVE A NEW JOB FOR YOU. THINK YOU'LL LIKE IT. CALL ME.

RESPONSE: THANK YOU BUT I'M QUITE HAPPY RIGHT HERE. HOW ARE YOU LIKING IT?

MEANT FOR: MY SISTER

EXPLANATION: I KNEW MY SISTER DIDN'T LIKE HER JOB, BUT I ACCIDENTALLY SENT IT TO MY OWN BOSS INSTEAD.

SENT TO: MY GRANDMOTHER

TXT MESSAGE: HEY HOT STUFF. ARE YOU COMING TO THE CITY THIS WEEKEND? KISSES

RESPONSE: I DON'T PLAN TO. WHO SAID THAT?

MEANT FOR: A GIRL I WAS DATING.

EXPLANATION: I DON'T USUALLY CALL MY GRANDMOTHER HOT STUFF...

SENT TO: MY BOSS

TXT MESSAGE: I WILL CALL IN SICK TODAY. NOT
SURE WHAT YOU DID TO ME LAST NIGHT. I CAME
TWICE. CAN YOU DO IT AGAIN TONIGHT? ;)

RESPONSE: YOU CAN STAY HOME TODAY, SOUNDS
LIKE YOU NEED IT. BUT TAKE IT EASY TONIGHT,
NEED YOU AT WORK TOMORROW.

MEANT FOR: MY HUSBAND

EXPLANATION: MY HEART STOPPED WHEN I READ
THE REPLY FROM MY BOSS.

SENT TO: GRANDPA

TXT MESSAGE: BECAUSE YOU HAVE BIG HANDS
AND FEET, DOES IT MEAN YOU HAVE SOMETHING
ELSE THAT'S BIG? ;)

RESPONSE: GLOVES AND SHOES. I DON'T THINK
YOU SHOULD SEND THINGS LIKE THIS TO BOYS.

MEANT FOR: A GUY FRIEND

EXPLANATION: MY GRAMPS APPARENTLY HAS
BEEN AROUND THE BLOCK.

SENT TO: DAD

TXT MESSAGE: CAN YOU PICK UP SOME BEER AND VODKA FOR ME? IF YOU DO I WON'T TELL MUM AND DAD THAT YOU USED THE CAR LAST WEEKEND? DEAL?

RESPONSE: I WANT TO SPEAK TO BOTH YOU AND PHIL TONIGHT!

MEANT FOR: MY BROTHER

EXPLANATION: IT WAS A LONG NIGHT, AND THEN MY BROTHER WAS UPSET THAT I DRAGGED HIM INTO THIS MESS.

SENT TO: MY MOTHER

TXT MESSAGE: IF YOU BUY SOME VASELINE ON YOUR WAY HOME WE CAN TRY IT IN NO 2 TONIGHT. YOU WANT TO?

RESPONSE: HONEY, JUST DON'T DO ANYTHING YOU ARE NOT COMFORTABLE WITH.

MEANT FOR: MY BOYFRIEND OBVIOUSLY

EXPLANATION: MY BOYFRIEND DIDN'T COME BY OUR HOUSE FOR WEEKS AFTER THIS.

SENT TO: MY DAD

TXT MESSAGE: TAG TEAMED BY KC AND JOJO LAST NIGHT. CAN HARDLY WALK TODAY BUT IT WAS WORTH IT.

RESPONSE: IS THIS HOW WE RAISED YOU?

MEANT FOR: KELLY, MY BEST FRIEND.

EXPLANATION: LESSON ONE: NEVER TALK TO YOUR PARENTS ABOUT BEING TAG TEAMED.

SENT TO: FRIEND LISA

TXT MESSAGE: MY USUAL BAD LUCK. JUST SENT A
TXT THAT LISA OUGHT TO LOSE SOME WEIGHT —
TO LISA! I MEANT TO SEND IT TO YOU.

RESPONSE: YOU HAVE SENT IT TO ME AGAIN, AND
MY FEELINGS ARE STILL HURT.

MEANT FOR: ANOTHER FRIEND...

EXPLANATION: I SENT HER THE WRONG TEXT TWICE
IN A ROW...I REALLY CAN'T EXPLAIN WHY.

SENT TO: ONE NIGHT STAND

TXT MESSAGE: IT WAS SICK, HER TITS POINTED IN TWO DIRECTIONS. I GOT DIZZY JUST FROM LOOKING AT THEM SO I TOOK HER FROM BEHIND INSTEAD.

RESPONSE: YOU ARE SICK, TITTY-BOY.

MEANT FOR: A FRIEND

EXPLANATION: NOT SURE IF SHE WAS OFFENDED OR TOOK THE WHOLE THING WITH EASE.

SENT TO: GRANDMOTHER

TXT MESSAGE: I CAN'T TAKE IT ANYMORE.
I'M DYING.

RESPONSE: HANG ON, DEAR.
I'M CALLING YOUR MOTHER.

MEANT FOR: MY FRIEND SARA

EXPLANATION: WE WERE ON THIS LONG, LONG HIKE
AND IT WAS SO WARM AND I TEXTED MY FRIEND
SARA, OR SO I THOUGHT... GRANDMA THOUGHT
SHE WOULD NEVER SEE ME AGAIN.

SENT TO: DAD

TXT MESSAGE: DAD WILL KILL ME IF HE FINDS OUT I USED THE CAR LAST WEEKEND. DON'T SAY ANYTHING TO ANYONE.

RESPONSE: I HEAR YOU KNOW ME WELL.

MEANT FOR: MY BROTHER

EXPLANATION: SOMEHOW I GOT OUT OF THIS EASY. I THINK MY DAD THOUGHT IT WAS VERY FUNNY.

SENT TO: EVA

TXT MESSAGE: WAS AT EVA'S PLACE YESTERDAY AND ACCIDENTALLY BROKE THAT ANTIQUE ASHTRAY SHE KEEPS ON HER BOOKSHELF. I SORT OF PUSHED IT TO THE SIDE AND SHE NEVER NOTICED. SHOULD I TELL HER?

RESPONSE: SHE KNOWS.

MEANT FOR: MY OTHER BEST FRIEND

EXPLANATION: I THINK GOD MADE ME TEXT EVA TO CONFESS MY SIN.

SENT TO: MY TEACHER

TXT MESSAGE: CALL ME IN THE MORNING TO MAKE SURE I GET UP. XOXO

RESPONSE: I THINK YOU SHOULD TAKE CARE OF THAT YOURSELF.

MEANT FOR: MY BOYFRIEND

EXPLANATION: MY TEACHER AND BOYFRIEND WERE RIGHT NEXT TO EACH OTHER IN MY CONTACTS LIST...

SENT TO: BOSS AT OLD JOB

TXT MESSAGE: I MET WITH DAPHNE AND SIGNED ON WITH THEM. TIME TO START THINKING ABOUT WHAT TO TELL MY OLD JOB SO I CAN LEAVE WITH MY HONOR INTACT.

RESPONSE: WHAT DO YOU MEAN!!!? I WANT TO SEE YOU IN MY OFFICE ASAP!!

MEANT FOR: A FRIEND

EXPLANATION: I ACTUALLY ENDED UP STAYING WITH MY OLD JOB, WHERE I GOT A BIG RAISE.

SENT TO: DADDY-O

TXT MESSAGE: MAYBE I SHOULD BE IN PORNOS.
THEY MAKE SICK MONEY. HOW HARD CAN IT BE?
SOME SUCK AND FUCK AND THEN YOU GO HOME.

RESPONSE: I'M WARNING YOU.

MEANT FOR: MY FRIEND ELLIE

EXPLANATION: I WAS JUST JOKING BECAUSE MY
FRIEND AND I HAD TALKED ABOUT HOW TO MAKE
MONEY.

SENT TO: MY FATHER

TXT MESSAGE: JUST HAVE TO TELL YOU, JOHN PUSHED HIS BALLS UP INTO HIS STOMACH YESTERDAY. THERE WAS JUST AN EMPTY SACK LEFT! A-HA-HA! CAN YOU DO THAT?

RESPONSE: I'M NOT SURE.

MEANT FOR: FRIEND SPENCER

EXPLANATION: THOUGHT I SENT IT TO MY FRIEND AND WHEN I REALIZED IT WENT TO MY DAD I WAS TOTALLY ASHAMED.

SENT TO: MY FATHER

TXT MESSAGE: I JUST HAVE TO ASK, HOW OFTEN DO YOU GUYS HAVE SEX?

RESPONSE: I PREFER NOT TO TALK TO YOU ABOUT THAT.

MEANT FOR: A FRIEND

EXPLANATION: I'M ACTUALLY VERY HAPPY I DIDN'T FIND OUT HOW OFTEN MY PARENTS HAVE SEX.

SENT TO: BOSS

TXT MESSAGE: I HAVE BUBBLES IN MY VAGINA. LIKE COKE BUBBLES. MAYBE IT FELL ASLEEP?

RESPONSE: TELL IT TO STAY AWAKE UNTIL OFFICE HOURS ARE OVER.

MEANT FOR: MY BEST GIRLFRIEND

EXPLANATION: I'D NEVER HAD THAT SENSATION BEFORE, SO I WAS SLIGHTLY WORRIED AND WANTED TO ASK MY FRIEND.

SENT TO: A GUY I WAS DATING

TXT MESSAGE: PSST. I'M IN HIS BATHROOM AND HAVE DONE A NUMBER 2, BUT THERE'S NO FUCKING TOILET PAPER. YOU HAVE TO GET OVER HERE.

RESPONSE: A KNOCK ON THE DOOR AND HE ASKED IF I NEEDED PAPER.

MEANT FOR: MY ROOMMATE

EXPLANATION: I WAS BEGINNING TO PANIC WHEN I FELT THE EMPTY ROLL AND SENT THE TEXT TOO QUICKLY....

SENT TO: MY DAD

TXT MESSAGE: GOT NO MORE CASH AND PAYDAY IS FAR AWAY. U WANT TO ROB A BANK?

RESPONSE: I WANT YOU TO COME HOME IMMEDIATELY. YOU BETTER HAVE A GOOD EXPLANATION FOR THIS.

MEANT FOR: MY BEST FRIEND ERIC

EXPLANATION: MY DAD THOUGHT I WAS SERIOUS, BUT I WAS OBVIOUSLY ONLY JOKING. BUT I KNEW I WOULDN'T GET CAUGHT IF I ROBBED A BANK.

SENT TO: DANIEL

TXT MESSAGE: DANIEL TEXTED ME THAT HE WAS GOING TO THE MOVIES WITH CAITLYN, AND I FEEL LIKE SHIT AND NOW I KNOW WHY, I'M IN LOVE WITH DANIEL!

RESPONSE: I CAN'T FREAKIN' BELIEVE THIS, I'M IN LOVE WITH YOU TOO!

MEANT FOR: SOMEONE ELSE BUT I DON'T CARE

EXPLANATION: THIS WAS THE BEST THING THAT EVER COULD HAVE HAPPENED. DESTINY REALLY BROUGHT US TOGETHER BY A STUPID MISTAKE.

SENT TO: GIRLFRIEND DAPHNE

TXT MESSAGE: LET DAPHNES CAT OUT, IT HASN'T COME BACK SINCE YESTERDAY EVENING. DAPHNE SAID IT WAS AN INDOOR CAT AND I PROMISED NOT TO LET IT OUT. WHAT SHOULD I DO? DO THEY COME BACK? CAN I BUY ONE JUST LIKE IT? I'M ABOUT TO FREAK OUT. SHE'S HOME TOMORROW. YOU GOTTA HELP ME.

RESPONSE: I'M HOPING THIS IS SOME KIND OF BAD JOKE.

MEANT FOR: A FRIEND

EXPLANATION: IT WAS FOR THE BEST BECAUSE WHEN SHE GOT HOME SHE WAS ALREADY PREPARED. THE CAT NEVER CAME HOME. DAPHNE STILL CALLS IT EVERY TIME SHE LEAVES HER HOUSE.

SENT TO: HEAD OF COMPANY

TXT MESSAGE: IF YOU ARE OUT RUNNING AROUND COULD YOU PICK UP A PACK OF TAMPONS FOR ME PLEASE? YOU ARE THE BEST.

RESPONSE: CAN'T, I'M ON MY WAY TO A MEETING. AND I THINK THIS IS SOMETHING YOU SHOULD BUY YOURSELF.

MEANT FOR: MY HUSBAND

EXPLANATION: I WAS SURE THIS WAS SENT TO MY HUSBAND. MY BOSS IS THE HEAD OF A MAJOR CORPORATION AND NOT SOMEONE YOU ASK TO RUN ERRANDS.

SENT TO: A ONE NIGHT STAND

TXT MESSAGE: NOW I KNOW WHY THERE IS SILICON. IT WAS LIKE SLEEPING ON A BOARD.

RESPONSE: ...AND YOU NEED A PENIS ENLARGEMENT. GOOGLE A CLINIC AND WE CAN GO TOGETHER.

MEANT FOR: MY BUDDY

EXPLANATION: MET A GIRL FOR A NIGHT BUT CAN'T SAY HOW I COULD BE SO STUPID TO SEND HER THE TEXT. AND MY PENIS ISN'T SMALL.

SENT TO: MY BOSS GREG

TXT MESSAGE: COULD YOU PLEASE GET IN FRONT OF THE WEBCAM TONIGHT AND SHAKE THINGS UP? SMOOCHES!

RESPONSE: SURE. ARE YOU DRUNK OR WHAT? YOU BETTER BE READY TOMORROW FOR THE MEETING.

MEANT FOR: A GIRL I WAS DATING.

EXPLANATION: MY BOSS IS SORT OF UPTIGHT BUT TOOK THIS UNUSUALLY WELL.

SENT TO: MY MOTHER

TXT MESSAGE: HA-HA, WE'LL SNEAK SOME VODKA AND JUST FILL THE BOTTLE BACK WITH WATER. THEY'LL NEVER KNOW. HA-HA.

RESPONSE: CHANGE OF PLANS. YOUR MOTHER AND I ARE STAYING HOME, AND SO ARE YOU. —DAD

MEANT FOR: MY FRIEND JESSICA

EXPLANATION: WE WERE PLANNING TO HAVE A PARTY BUT INSTEAD I HAD TO SIT THROUGH A PRETTY SCARY DINNER WITH MY PARENTS.

SENT TO: MY BOSS

TXT MESSAGE: TODAY MY FARTS DON'T REALLY SMELL BAD. QUITE THE OPPOSITE, THEY ACTUALLY SMELL PRETTY GOOD.

RESPONSE: PLEASE, I WILL NOT RESPOND TO THIS.

MEANT FOR: MY HUSBAND

EXPLANATION: WE USUALLY SEND EACH OTHER WEIRD TEXTS, BUT WE USUALLY DON'T LET MY BOSS IN ON IT...

SENT TO: A GIRL I JUST MET

TXT MESSAGE: I MUST HAVE MESSED WITH MY PROSTATE OR SOMETHING WHEN I TOOK A DUMP, BECAUSE I GOT A FEVER SHORTLY AFTER. WHAT THE EFF?

RESPONSE: YEAH, WHAT THE EFF?!!!! YUCK

MEANT FOR: MY FRIEND TOM

EXPLANATION: TOM LAUGHED SO HARD HE ALMOST CRAPPED IN HIS PANTS WHEN HE HEARD THIS, AND I WAS DEVASTATED.

SENT TO: HOT GIRL GILLIAN

TXT MESSAGE: DUDE, YOU'VE GOTTA DO ME A HUGE FAVOUR. STAYING WITH GILLIAN TONIGHT BUT HAVEN'T CHANGED UNDERWEAR SINCE YESTERDAY. HAVEN'T CHECKED BUT I KNOW IT'S NOT GOOD. CAN YOU COME BY WITH A FRESH PAIR? CALL WHEN OUTSIDE AND I'LL MEET YOU. I'LL MAKE UP SOMETHING.

RESPONSE: AAAAAHHH-HA-HA-HA! YOU SENT THIS TO ME!

MEANT FOR: MY ROOMMATE

EXPLANATION: I CAN'T THINK OF ANYTHING MORE EMBARRASSING HAPPENING TO ME.

SENT TO: TEACHER MR. HARRIS

TXT MESSAGE: JEEEEEZ! MY ONE BOOB IS BIGGER THAN THE OTHER. HELP! DOES THIS MEAN I WILL GET A HANG-TIT? LOOKS WEIRD!

RESPONSE: I DON'T THINK THIS IS DANGEROUS BUT JUST TO MAKE SURE YOU SHOULD VISIT THE NURSE. THANK YOU FOR YOUR CONFIDENCE IN ME.

MEANT FOR: OBVIOUSLY NOT HIM.

EXPLANATION: I CAN'T GO TO CLASS WITHOUT BEING REMINDED OF THIS. NOW I KNOW HE KNOWS WHAT MY BOOBS LOOK LIKE AND I WISH I WOULD JUST GRADUATE TOMORROW...

SENT TO: MY GRANDFATHER

TXT MESSAGE: HOW ARE YOU FEELING TODAY, HEARD YOU SCORED LAST NIGHT?

RESPONSE: WHAT DO YOU MEAN?

MEANT FOR: MY FRIEND JOSH

EXPLANATION: GRAMPS STILL TALKS ABOUT THIS AND THINKS I THINK HE'S A STUD OR SOMETHING. I LET HIM BELIEVE THAT I DO.

SENT TO: MY MUM

TXT MESSAGE: CAN'T TALK NOW, TOO DRUNK. CALL 2MORROW.

RESPONSE: WHERE ARE YOU? YOUR FATHER WILL COME PICK YOU UP IMMEDIATELY. WE GET REALLY WORRIED WHEN YOU SAY THINGS LIKE THAT.

MEANT FOR: MY BUDDY

EXPLANATION: WHAT CAN I SAY — IT WAS TRUE. MY DRUNK ASS SENT THE TEXT TO THE WRONG PERSON.

SENT TO: MY FATHER

TXT MESSAGE: I WAS ONLY PRETENDING TO
SLEEP AND WATCHED YOU RUN AROUND
NAKED. ME LIKE ;)

RESPONSE: THIS IS VERY STRANGE FOR ME
TO HEAR. LETS NOT TALK ABOUT THIS AGAIN.

MEANT FOR: MY BOYFRIEND

EXPLANATION: MY BOYFRIEND SLEPT OVER BUT
GOT UP EARLY SO MY PARENTS WOULDN'T NOTICE.
NOW THEY KNOW SINCE I HAD TO EXPLAIN IT TO
MY MOTHER.

SENT TO: CATHY

TXT MESSAGE: HOLY SMOKES, IT SEEMS AS IF CATHY IS FINALLY PREGNANT. I'M HAPPY FOR THEM. I KNOW THEY'VE BEEN TRYING FOR A WHILE.

RESPONSE: I'M SORRY, I GUESS THAT MEANS I'VE GAINED SOME WEIGHT THEN:/

MEANT FOR: OUR MUTUAL FRIEND SARA.

EXPLANATION: I'D VISITED MY FRIEND CATHY WHOM I KNOW HAD TRIED TO GET PREGNANT FOR MONTHS. I WAS POSITIVE THAT ROUND BELLY OF HERS WAS A PREGGO BELLY, BUT THAT ONLY PROVES HOW MUCH I KNOW...

SENT TO: MY GRANDFATHER

TXT MESSAGE: YO MAN! YOU JOINING THE PARTY
TONIGHT? IT WILL BE SICK. I GOT THE DRINKS.

RESPONSE: NO RESPONSE

MEANT FOR: MY FRIEND

EXPLANATION: I NEVER HEARD ANYTHING ABOUT
THIS, PERHAPS HE NEVER EVEN GOT IT, OR HE
CHOSE NOT TO MENTION IT. IT WOULD HAVE BEEN
COOL IF HE'D COME ALONG THOUGH.

SENT TO: GRANNIE

TXT MESSAGE: YOU COMING TO SHOOT SOME HOOPS TODAY?

RESPONSE: I DON'T THINK I CAN. MY KNEE IS STILL BOTHERING ME.

MEANT FOR: MY B-BALL FRIEND

EXPLANATION: MY POOR OLD GRANDMA GOT CON-FUSED WHEN I ACCIDENTALLY ASKE D HER IF SHE WANTED TO SHOOT SOME HOOPS. I HAD A BAD DAY THOUGH, SO SHE PROBABLY WOULD HAVE BEAT ME TOO...

SENT TO: THE BOSS

TXT MESSAGE: I HAD A DREAM THAT I BEAT YOU UP LAST NIGHT. I'M SURE YOU HAD IT COMING. PERHAPS I'LL KNOCK YOU OUT WHEN I SEE YOU ;) XO

RESPONSE: OK, YOU ARE SCARING ME.

MEANT FOR: THE BOYFRIEND

EXPLANATION: WAS OBVIOUSLY MEANT FOR MY BOYFRIEND. PERHAPS NOT THE WORST I COULD HAVE SAID BUT IT WAS STILL EMBARRASSING.

SENT TO: JOE

TXT MESSAGE: HOLY SHIT. SLEPT AT JOE'S LAST NIGHT AND HE LET ONE RIP IN THE MIDDLE OF THE NIGHT. IT WOKE ME UP. I DON'T THINK I CAN SEE HIM ANYMORE.

RESPONSE: OOPS... DON'T KNOW WHAT TO SAY

MEANT FOR: MY FRIEND

EXPLANATION: I'M AN IDIOT BUT I'D BEEN GAS POISONED ALL NIGHT SO COULDN'T THINK STRAIGHT.

SENT TO: MY FATHER

TXT MESSAGE: DO YOU WANT TO HELP ME SHAVE
IN THE SHOWER TONIGHT ;)

RESPONSE: NO RESPONSE

MEANT FOR: MY FUTURE HUSBAND

EXPLANATION: MY DAD HAD ENOUGH TACT NEVER
TO TALK ABOUT THIS.

SENT TO: BOSS

TXT MESSAGE: I'M GOING TO BED NOW.

RESPONSE: OK. WELL, GOOD NIGHT.

MEANT FOR: A GUY I WAS DATING.

EXPLANATION: HE WAS SUPPOSED TO COME
OVER BUT WAS LATE AND I TEXTED MY BOSS
BY MISTAKE.

SENT TO: DAD

TXT MESSAGE: HURRY, I'M WARMING UP THE BED!

RESPONSE: NO. NOT WHILE YOUR MOTHER AND I ARE IN THE HOUSE.

MEANT FOR: MY BOYFRIEND

EXPLANATION: BY MISTAKE I TEXTED MY DAD IN-STEAD OF MY BOYFRIEND. THE NEXT MORNING WAS THE WORST OF MY LIFE. I STAYED IN BED UNTIL NOON.

SENT TO: FRIEND MIRIAM

TXT MESSAGE: HONESTLY, IS SOMETHING
WRONG WITH MIRIAM'S BABY? IT LOOKS SORT
OF MONGOLOID.

RESPONSE: AND YOU CALL YOURSELF MY
FRIEND???????

MEANT FOR: ANOTHER FRIEND...

EXPLANATION: WE'VE NEVER TALKED ABOUT THIS
EVEN THOUGH WE STILL SEE EACH OTHER.

SENT TO: MY BOSS

TXT MESSAGE: I JUST SAW A MOVIE THAT CURED ME OF MY FEAR OF DEATH!

RESPONSE: GOOD FOR YOU!

MEANT FOR: A FRIEND

EXPLANATION: WAS JUST MESSING AROUND WITH A FRIEND BUT MY BOSS SOMEHOW GOT IT INSTEAD.

SENT TO: JIM, FRIEND WITH BENEFITS

TXT MESSAGE: YO GIRL. SLEPT AS JIM'S LAST NIGHT. SEX GOOD AS USUAL BUT HE FARTED SEVERAL TIMES IN HIS SLEEP. I DIDN'T KNOW WHAT DO TO. HE WOULD DIE IF HE KNEW. CALL YOU LATER. CIAO.

RESPONSE: OK, NOW I KNOW. AND YOU ARE RIGHT, I FEEL LIKE DYING. BUT EVEN MORE EMBARRASSING THAT YOU TELL YOUR FRIENDS. CAN'T YOU JUST LEAVE IT ALONE?

MEANT FOR: MY GIRL FRIEND

EXPLANATION: I DON'T KNOW WHAT I WAS THINKING TEXTING HIM.

SENT TO: MUM

TXT MESSAGE: THAT'S IT, I'VE DECIDED TO BECOME A LESBIAN!

RESPONSE: HONEY, I'VE ALWAYS SUSPECTED THIS BUT CAN'T YOU COME OVER SO WE CAN TALK ABOUT THIS FACE TO FACE? KISSES

MEANT FOR: MY FRIEND

EXPLANATION: I WAS JUST VENTING MY BAD BOY-FRIEND EXPERIENCE TO A FRIEND, I THOUGHT. THAT MY MOTHER SUSPECTED I WAS A LESBIAN WAS QUITE SHOCKING...

SENT TO: GRANDMA

TXT MESSAGE: I WON'T BE ABLE TO GO THIS YEAR. MY DAD SAID THAT GRANDPA WILL PROBABLY NOT LIVE THAT LONG AND I WANT TO BE HOME WHEN THAT HAPPENS.

RESPONSE: NO RESPONSE

MEANT FOR: I MEANT TO SEND THIS TO MY FRIEND WHO I TALKED TO ABOUT GOING TO A SKATEBOARD CAMP.

EXPLANATION: I DON'T THINK MY GRANDMA OR GRANDPA EVER READ IT BECAUSE LUCKILY THEY DON'T KNOW HOW TO WORK THEIR CELL PHONE PROPERLY.

SENT TO: ONE NIGHT STAND

TXT MESSAGE: I TOLD HER I HAD TO GO TO THE OFFICE SO SHE JUST LEFT. U IMAGINE, WOBBLING HOME ON HIGH HEELS, SMEARED MAKE-UP AND A WRINKLY COCKTAIL DRESS? YOU KNOW EVERY-ONE THAT SEES YOU KNOWS YOU'RE ON A WALK OF SHAME. HA-HA! PIZZA IN AN HOUR?

RESPONSE: YOU ARE SO PATHETIC I WON'T EVEN RESPOND TO YOUR TXT.

MEANT FOR: MY BEST FRIEND

EXPLANATION: I SORT OF FELT LIKE A JERK AFTER THIS. WE NEVER MET AGAIN.

SENT TO: MUM

TXT MESSAGE: BRING AN EXTRA KNIFE, IN CASE WE BOTH HAVE TO CHOP OFF A HEAD.

RESPONSE: STOP THIS, HOW CAN YOU SAY SOMETHING LIKE THAT. ANSWER WHEN I CALL YOU.

MEANT FOR: MY BUDDY STEVE

EXPLANATION: WE WERE GOING FISHING.
MY MOTHER THOUGHT WE WERE ABOUT
TO KILL SOMEONE…

SENT TO: DAD

TXT MESSAGE: WHAT WAS THAT WEBSITE ABOUT
BOOB JOBS?

RESPONSE: I THINK THAT YOU, YOUR MOTHER, AND
I SHOULD TALK ABOUT THIS BEFORE YOU EVEN
VISIT THAT WEBSITE.

MEANT FOR: MY FRIEND

EXPLANATION: UNFORTUNATELY I SENT THIS TO
MY DAD INSTEAD OF MY FRIEND, AND UNFORTU-
NATELY I STILL HAVE SMALL BOOBS...

SENT TO: MY BOSS

TXT MESSAGE: I'M STILL SORE FROM LAST NIGHT. YOU WERE A COMPLETE BEAST AND TOOK ME SO HARD I'M HAVING TROUBLE WALKING TO WORK ☺ SEE YOU TONIGHT BEAST XX

RESPONSE: YOU CAN DO IT. ONE STEP AT A TIME.

MEANT FOR: MY HUSBAND

EXPLANATION: SENT THIS TO MY BOSS INSTEAD OF MY HUSBAND. I WAS THINKING ABOUT QUITTING.

SENT TO: MY BROTHER

TXT MESSAGE: HA-HA, HOME ALONE AND I'VE JUST TRIED THE NEW RABBIT VIBRATOR. YOU WERE RIGHT, IT WAS AWESOME! ☺

RESPONSE: I WILL IGNORE THIS TEXT, REMOVE IT FROM MY INBOX AND PRETEND IT NEVER HAPPENED

MEANT FOR: MY FRIEND WHO GAVE ME A VIBRATOR FOR CHRISTMAS...

EXPLANATION: I COULD NOT LOOK MY BROTHER IN THE FACE FOR MONTHS AFTER THIS.

SENT TO: BENJAMIN

TXT MESSAGE: LAST NIGHT WHEN I WAS WITH BEN I ACCIDENTLY FARTED. I TOLD HIM IT WAS A PUSSY FART. I THOUGHT I WAS GOING TO DIEEEEEEEEE!!!

RESPONSE: WHAT THE F?

MEANT FOR: MY BEST FRIEND CHARLY

EXPLANATION: I SENT THIS TO THE GUY I FARTED ON INSTEAD OF MY FRIEND. IF I COULD HAVE DISAPPEARED AFTER I GOT HIS TEXT I WOULD HAVE.

SENT TO: GIRLFRIEND MIA

TXT MESSAGE: I'M THINKING OF BREAKING UP WITH MIA. NOT SURE IT WORKS BETWEEN US, AND I'M NOT REALLY IN LOVE ANYMORE. WHAT DO YOU THINK I SHOULD DO?

RESPONSE: ARE YOU JOKING?

MEANT FOR: ONE OF OUR MUTUAL FRIENDS

EXPLANATION: OBVIOUSLY THIS WASN'T MEANT FOR MY GIRLFRIEND...

SENT TO: MY FEMALE BOSS

TXT MESSAGE: WHAT FEELS BEST, TAKING A CRAP WHEN YOU REALLY HAVE TO GO, OR HAVING SEX WHEN YOU ARE REALLY HORNY?

RESPONSE: EXCUSE ME?

MEANT FOR: A BUDDY

EXPLANATION: MY FRIEND WAS DESPERATE TO TAKE A CRAP BUT WAS STUCK ON THE BUS WHEN HE TEXTED ME, BUT FOR SOME REASON I SENT THIS TO MY BOSS INSTEAD.

SENT TO: FRIEND SUSIE

TXT MESSAGE: DON'T YOU THINK SUSIE'S BABY LOOKED SORT OF WEIRD? A FLAT NOSE AND BIG EYES? SORT OF EXTRA TERRESTRIAL.

RESPONSE: YOU'VE SENT THIS TO ME. LOOKING FORWARD SEEING WHAT YOUR BABY WILL LOOK LIKE, CONSIDERING HOW UGLY YOU ARE.

MEANT FOR: MY FRIEND SUSANNA

EXPLANATION: WELL, WHAT CAN I SAY, THE BABY WAS UGLY BUT I GUESS NO PARENT WANTS TO HEAR THAT.

SENT TO: DAD

TXT MESSAGE: MAN, IT SMELLS LIKE LUCKY CHARMS
WHEN I TAKE A PISS. IS SOMETHING WRONG?
U THINK I SHOULD GET TESTED?

RESPONSE: HA-HA, YOU'VE HAD BEER.
I KNOW THAT SMELL.

MEANT FOR: A FRIEND.

EXPLANATION: WAS MEANT FOR A FRIEND. GOT
WORRIED I HAD SOME TROLLS IN MY FLUTE, OR
SOMETHING, BUT MY DAD HAD OBVIOUSLY BEEN
THERE.

JUST AS GOOD AS CONDOMS

YO. THIS IS ANA FROM LAST THURSDAY. WAS JUST THINKING, ARE YOU SURE YOU DON'T HAVE ANY DISEASES? HIV AND ALL? HUGZ

BIKING FISH ARE USELESS

A GIRL NEEDS A GUY AS MUCH AS SHE NEEDS A BIKE. I MEAN, AS A FISH NEEDS A BIKE. EHHM, YOU KNOW WHAT I MEAN.

IN THE JOB DESCRIPTION

I HATE IT WHEN CHICKS WANT CREDIT FOR BLOWING YOU. WHAT THE HELL, THAT'S PART OF BEING A CHICK TO BEGIN WITH.

DEEP SEA LOVE

IT FEELS LIKE I WAS RAPED BY A WALRUS. NO MATTER WHAT I CAN'T SEEM TO GET THE SCENT OF FISH OFF, AND MY COCK IS ALL SCRATCHED.

RANDOM "BURP!" DRUNKEN "BURP!" TEXTS FROM LATE, LATE NIGHTS:

NERVE WRACKING

IF THEY EVER COME UP WITH THE PILL FOR US MEN I'D STAND IN LINE. I CAN'T TAKE ANOTHER FUCKING TIME HEARING THAT HER PERIOD IS LATE.

WILD LOVE

BANG, BANG AND THE BEARS ARE DEAD! I'M A SLUT AND MY BED REEKS OF MEN.

A TRUE GENTLEMAN

YO GIRL, YOU MAY HAVE HERPES.

COMMUTER LOVE

PLEASE SHOOT ME. AT GRAND CENTRAL LOOK-ING LIKE A RUSSIAN WHORE WITH WET HAIR. BEEN SCREWING FOR LIKE 3 HOURS. I'M DEAD. XX

HO, HO, HO

I WILL START ON THE ICE CREAM IF I DON'T ASK. DID WE USE A CONDOM THIS WEEKEND? JUST WANT TO MAKE SURE I'M NOT ROLLING THROUGH THE STREETS COME CHRISTMAS.

IN THE GRIPS OF DESIRE

I GET A HARD-ON IF SHE JUST LOGS IN TO MSN.

RANDOM "BURP!" DRUNKEN "BURP!" TEXTS FROM LATE, LATE NIGHTS:

EXTRA EVERYTHING PLEASE

IT TASTED LIKE HE'D RUBBED HIS COCK IN
ONIONS. WOULD YOU SUCK THAT??!!

SMOOTH IS FOR BABIES

HOW THE HELL CAN SOMEONE DUMP YOU FOR
SHAVING YOUR BALLS??

TROUBLE IN PARADISE

DO YOU KNOW IF THERE'S A QUICK TEST FOR CHLA-MYDIA HERE IN THAILAND? GOTTA TAKE CARE OF THIS BEFORE MARIA GETS HERE.

BRUTAL TRUTH

HE MUST HAVE THOUGHT I WAS BLIND. DON'T KNOW WHY HE EVEN BOTHERED TALKING TO ME.

RANDOM "BURP!" DRUNKEN "BURP!" TEXTS FROM LATE, LATE NIGHTS:

CHOOSE AND YOU CAN'T LOOSE

IF I CAN'T TIE YOU UP YOU CAN'T FUCK ME.
SIMPLE AS THAT.

WHAT HAPPENS IN VEGAS, YOU FORGET...

OOOPS, I MAY HAVE GOTTEN MARRIED...

FOR THE GOOD OF MANKIND

I'M GOING INTO GENE SCIENCE TO MAKE SURE THAT FUTURE MEN HAVE A TONGUE RIGHT ABOVE THEIR PENIS!

BAD NEWS THEN GOOD NEWS

WHEN I FINISHED LAUGHING I GAVE HIM A BLOWJOB.

RANDOM "BURP!" DRUNKEN "BURP!" TEXTS FROM LATE, LATE NIGHTS:

THAT'S SO EUROPE

THIS IS WHAT HAPPENED. I'M AT THE BEACH WITH ONE HUNDRED THOUSAND OTHER PEOPLE AND IT'S THE MIDDLE OF THE NIGHT. I TALK TO AN OK LOOKING GUY WITH A PONYTAIL WHO'S FROM HOLLAND AND LEAVE WITH HIM. HE PAYS FOR THE CAB AND CUMS BEFORE WE HAVE SEX. I FALL ASLEEP AND WHEN I WAKE UP HE IS ON A MATTRESS ON THE FLOOR. THE SECOND TIME I WAKE THE GUY SAYS THAT HIS LANDLADY WILL BE THERE IN 20 TO CHECK THAT HE HASN'T GOT ANYONE IN HIS ROOM. IT'S SEVEN IN THE MORNING AND HE GIVES ME 10 EURO FOR A TAXI. THE FAIR IS ONLY 5 SO HERE I AM, JUST MADE MYSELF 5 EURO TO LET AN UGLY, DUTCH GUY CUM ON MY NEW DRESS. CALL ME LATER?

MASTER OF SIMILIES

DRINKING NON-ALCOHOLIC BEER IS LIKE GOING DOWN ON YOUR SISTER. IT TASTES PRETTY MUCH THE SAME BUT IT IS JUST SOOO WRONG.

SPONSORED BY KALUHA

IF I DRINK KALUHA I GET HOMOEROTIC DREAMS.

THE UNOFFICIAL SLOGAN

PUSSY WORKS LIKE MASTERCARD,
EVERYWHERE AND ALL THE TIME.

RANDOM "BURP!" DRUNKEN "BURP!" TEXTS FROM LATE, LATE NIGHTS:

BRA NOSTALGIA

HELL, YOU REMEMBER WHEN WE WERE YOUNGER AND IT WAS SORT OF CUTE THAT THE GUY COULDN'T GET THE BRA OFF? THE ONE FROM LAST NIGHT TOOK IT OFF AND I DIDN'T EVEN KNOW IT. MAKES ME FEEL SORT OF SAD.

FACEBOOK SCHMACEBOOK

I FELT THINGS WENT A BIT TOO FAR WHEN SHE SAID SHE HAD TO UPDATE HER FACEBOOK STATUS AFTER WE HAD SEX.

BUT IT'S MY FAVORITE SHOW

BUT REALLY. HE ASKED ME TO TURN THE TV ON WHEN I WAS BLOWING HIM. THEN HE STOPPED ME AGAIN AND ASKED ME TO HAND HIM THE REMOTE. I'M GOING HOME NOW.

MAKES YOU CURIOUS

MAN I HATE WHEN PEOPLE DO THAT. I HOPE YOU WAKE UP PISS DRUNK NEXT TO SOME GODDAMN HIPPO CHICK WITH A BUSTED CONDOM ON YOUR BELLY. GO TO HELL.

RANDOM "BURP!" DRUNKEN "BURP!" TEXTS FROM LATE, LATE NIGHTS:

CUTE PUSSY

JUST MET THIS GUY WHO ASKED ME HOME TO LOOK AT HIS CAT. THOUGHT HE WANTED TO FUCK ME BUT WE REALLY JUST LOOKED AT HIS CAT. WHERE ARE YOU?

NOW YOU SEE IT, NOW YOU DON'T

HE CAME SO FAST I DIDN'T EVEN HAVE TIME TO PREPARE FOR MY FAKE ORGASM.

HAIRY, SCARY

HAIR EVERYWHERE! PANIC, NEED A RIDE.

JUST ONE MORE SONG, DARLING

HE WANTED ME TO BLOW HIM WHILE HE PLAYED GUITAR HERO. NOT SURE THERE'S GOING TO BE A SECOND DATE.

RANDOM "BURP!" DRUNKEN "BURP!" TEXTS FROM LATE, LATE NIGHTS:

ROCKING CHAIR

THOUGHT IT WOULD BE FUN TO CALL HER GRANDMA WHILE SHE WAS ON TOP OF ME. I'M TELLING YOU, THAT JOKE WAS WASTED.

JUST GOT OFF THE BOAT

JUST DID A WALK OF SHAME FROM MANHATTAN TO BROOKLYN. TORN PANTYHOSE AND TRASHY HAIR. I LOOK LIKE A REFUGEE.

RANDOM "BURP!" DRUNKEN "BURP!" TEXTS FROM LATE, LATE NIGHTS:

WILL WORK FOR FOOD

I SHOWED A BOOB FOR A BEER. ARE YOU ASHAMED OF ME?

CAN'T BARS HAVE DECENT LIGHT?

NO WAY HE LOOKED LIKE THIS LAST NIGHT. I'D RATHER CHEW MY ARM OFF THEN WAKE HIM.

RANDOM "BURP!" DRUNKEN "BURP!" TEXTS FROM LATE, LATE NIGHTS:

MEASUREMENT OF A GOOD EVENING

FUCKED MY PIERCING OFF THE OTHER NIGHT.
WHAT A RIDE!

THE PRICE OF GREED

ALRIGHT, MY JAW IS REALLY SORE. I KNOW HE WAS
BIG BUT WTF!!?

RANDOM "BURP!" DRUNKEN "BURP!" TEXTS FROM LATE, LATE NIGHTS:

PARTY TRICK

IT ENDED WITH ME PUTTING PINEAPPLE RINGS AROUND MY PENIS. SOME WERE IMPRESSED BUT NOT IN THE GOOD WAY.

BAD TIMING

GREAT THAT I GOT MY PERIOD JUST AS HE PUT IT IN. IT WAS LIKE THE FUCKING NIAGARA FALLS. SIGH, LET'S CROSS THAT GUY FROM THE LIST AS WELL.

NIGHT WITH THE ALIENS

I KID YOU NOT, THERE WAS CUM EVERYWHERE. WE'RE TALKING THE WALLS, CLOTHES, MY FUCKING HAIR! IT WAS LIKE A SPLATTER MOVIE BUT WHITE GOO INSTEAD OF BLOOD. CREEEPY.

THE MILF TRAP

MILFS ARE GREAT FUN UNTIL YOU ARE FORCED TO EAT LUCKY CHARMS WITH HER KID THE MORNING AFTER.

RANDOM "BURP!" DRUNKEN "BURP!" TEXTS FROM LATE, LATE NIGHTS:

SNOOP DOG

EHHMM, JUST WOKE UP FROM HIS DOG BREATHING IN MY EAR. FIRST I THOUGHT IT WAS HIM BECAUSE IT SOUNDED EXACTLY LIKE HOW HE SOUNDED LAST NIGHT. THAT'S WHAT I CALL DOGGY STYLE ;)

STYLE AND EFFICIENCY

STOP TEXTING ME IF YOU DON'T WANT TO HAVE SEX. WON'T BOTHER READING THE CRAP.

RANDOM "BURP!" DRUNKEN "BURP!" TEXTS FROM LATE, LATE NIGHTS:

RODEO

A PAINTING CRASHED TO THE FLOOR, A LAMP FELL OVER AND HE STARTED TO BLEED FROM THE NOSE. GOOD SEX? HELLYEAH!

HE'S ONLY WATCHING

FUCKING DISTURBING THAT SHE HAS THIS HUGE PICTURE OF HER FATHER RIGHT ABOVE THE BED. WHO'S FACE SHOULD I LOOK AT?

ANYTHING CAN HAPPEN IN EUROPE

I LOVE ROMANIA. JUST MADE OUT WITH A DWARF.